HOW TO LEARN

THE 10 PRINCIPLES OF EFFECTIVE REVISION AND PRACTICE

Also by Fiona McPherson

Easy Russian Alphabet: A Visual Workbook

Indo-European Cognate Dictionary

Mnemonics for Study (2nd ed.)

My Memory Journal

Successful Learning Simplified: A Visual Guide

How to Approach Learning: What teachers and students
  should know about succeeding in school

Effective Notetaking (3rd ed.)

Planning to Remember: How to remember what you're
  doing and what you plan to do

Perfect Memory Training

The Memory Key

How to Learn

THE 10 PRINCIPLES OF EFFECTIVE REVISION AND PRACTICE

Fiona McPherson, PhD

Wayz Press
Wellington, New Zealand

Published 2018 by Wayz Press, Wellington, New Zealand.

ISBN 978-1-927166-13-0

To report errors, please email errata@wayz.co.nz

For additional resources and up-to-date information about any
errors, go to the Mempowered website at
www.mempowered.com

Contents

Figures

Tables & Lists

What you need to know about memory

To understand how to practice or revise effectively, you
need to understand some basic principles about how
memory works. This chapter covers:

- the 8 basic principles of memory
- how neurons work
- what working memory is and why it's so important
- what consolidation is and why it matters

'Practice' (a term I use to cover both revision of information
and the practice of skills) is a deceptively simple concept.
Everyone thinks they know what it means, and what they think
it is, basically, is repetition. When we have to remember an
unknown phone number long enough to dial it, we repeat it to
ourselves. As children told to learn a poem, we repeated it until
we had it memorized. Learning to drive, we repeated the
necessary actions over and over again. Repetition is at the heart
of learning.

But simple repetition is the least effective learning strategy there is.

How do we reconcile these statements? Well, repetition is crucial to cement a memory, but the untutored way of doing it wastes a great deal of time, and still results in learning that is less durable than more efficient strategies.

In this book, I'm going to tell you the 10 principles of effective repetition, why they work, and how they work. We'll look at examples from science, mathematics, history, foreign language learning, and skill learning. At the end of it, you'll know how to apply these principles in your study and your daily life.

Let's start with a very brief look at how memory works.

The 8 basic principles of memory

The most fundamental thing you need to understand about memory is that it is not a recording. When we put information into our memory, we don't somehow copy the real-world event, as a video camera might, but rather we select and edit the information. For this reason, putting information into memory is called **encoding**. This is why I habitually talk about memory codes rather than memories. It's a reminder that nothing in your memory is a 'pure' rendition, a faithful copy. We create our memory codes, and when we try and retrieve a memory, it is this coded information that we are looking for. (Like a computer, what our brain processes is information — when I use the word 'information', I don't just mean 'facts', but images and skills and events and everything else we file in long-term memory.)

Why does it matter that the information is coded?

Because what you think you are looking for may not be

precisely what is there. How easy it is to remember something (retrieve a memory code) depends on the extent to which the code matches what you think you're looking for.

For example, say you are trying to remember someone's name. You might think it begins with T, or that it's unusual, or very common, or sounds something like -immy, or that it's old-fashioned, or ... Whatever your idea is, the point is that there *is* an idea, a starting point, a clue (we call it a **retrieval cue**). How likely you are to retrieve the memory code depends on how good a clue it is.

This is because memory codes are linked together in a network. Remembering is about following a trail through the network, following the links. No surprise then that your starting point (the retrieval cue) is crucial.

For example, consider this simplified memory code for Henry VIII:

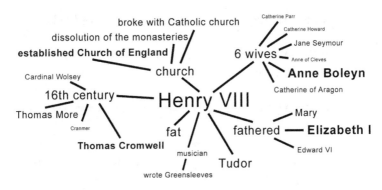

The size of the words reflect how strong those parts of the code are — Anne Boleyn, for example, is for most of us the most memorable of Henry's wives; Elizabeth the most memorable of his children.

Accordingly, it would be a lot easier to retrieve "Henry VIII" if the retrieval cue was "father of Elizabeth I" than if it was "father of Edward VI", or if the cue was "established the Church of England" rather than "Cranmer's king", or, worst of all, "16th century musician". (Do note that information in a memory code is not necessarily true! For example, Henry VIII did not actually write the song *Greensleeves*, but it is a common belief that he did. 'Information' is a blanket word to cover a type of content; the statement "The grass is green" and the statement "The grass is red" contain the same amount of information, although only one of the statements is true.)

The trail through memory resembles a trail through a jungle. Much-travelled paths will be easier and quicker to follow. Paths that have been used recently will be easier to find than old disused trails.

There are eight fundamental principles encompassed in these simple ideas:

1. **code principle**: memories are selected and edited codes.

2. **network principle**: memory consists of links between associated codes.

3. **domino principle**: the activation of one code triggers connected codes.

4. **recency effect**: a recently retrieved code will be more easily found.
   If you were watching the TV program *The Tudors* last night, it would be much easier to call up Henry VIII's name up again than it would be if you hadn't thought of him since school.

5. **priming effect**: a code will be more easily found if linked

codes have just been retrieved.

Having been thinking of Henry VIII, you will find it easier to retrieve "Walter Raleigh" (linked to Elizabeth I), compared to a situation where you were asked, out of the blue, who that guy was who put his cloak across the puddle for Queen Elizabeth to walk over.

6. **frequency** (or **repetition**) **effect**: the more often a code has been retrieved, the easier it becomes to find.

7. **matching effect**: a code will be more easily found the more closely the retrieval cue matches the code.
This can be seen in jokes: if you were asked, "What did the tree do when the bank closed?", you'd probably realize instantly that the answer had something to do with "branch", because 'branch' is likely to be a strong part of both your "tree" code and your "bank" code. On the other hand, if you were asked, "What tree is made of stone?", the answer (lime tree) is not nearly as easily retrieved, because "lime" is probably not a strong part of either your "tree" code or your "stone" code.

8. **context effect**: a code will be more easily found if the encoding and retrieval contexts match.
If you learned about Henry VIII from watching *The Tudors* on TV, you will find it easier to remember facts about Henry VIII when you're sitting watching TV. We use this principle whenever we try and remember an event by imagining ourselves in the place where the event happened.

These principles all affect how practice works and what makes it effective, but three are especially important. The recency and priming effects remind us that it's much easier to follow a memory trail that has been activated recently, but that's not a strength that lasts. Making a memory trail permanently stronger requires repetition (the frequency effect). This is about neurobiology: every time neurons fire in a particular sequence (which is what happens when you 'activate' a memory code), it makes it a little easier for them to fire in that way again.

The frequency effect is at the heart of why practice is so important. The recency and priming effects are at the heart of why most people don't practice effectively.

How neurons work

Let's take a very brief look at this business of neurons firing. Neurons are specialized brain cells. We might think of them as nodes in the network. Neurons are connected to each other through long filaments, one long one (the **axon**) and many very short ones (**dendrites**). It is the long axon that carries the outgoing signal from the neuron. The dendrites receive the incoming signals, and they do this through specialized receptors called **synapses**. For here's the thing: neurons aren't physically connected. Messages are carried through the network by electrical impulses along the filaments, which induce chemical responses at the synapses. Specialized chemicals called **neurotransmitters** travel the very short gap between the synapses on one neuron to those on a nearby one.

In other words, information is carried within the neuron in the form of electrical impulses (as it is in your radio and television), is then transformed into a chemical format so that it can cross the gap between neurons, and then translated back into electrical impulses in the receiving neuron.

Being carried as an electrical signal has an important

implication: how fast we think (and how well, as we'll see in the next section) depends on how fast the signals are flowing. The speed of the electrical signal depends on the wiring. As with the wiring in your home, the 'wires' (axons) are sheathed in insulation. The better insulated, the less 'loss' in the signal, the faster the signal can travel. In the brain, this insulation is called **myelin**. Because myelin is white(ish), and the cell bodies are gray, we commonly refer to 'gray matter' and 'white matter'.

Myelin tends to degrade over time, and this degradation is one of the factors implicated in cognitive decline in old age. Myelin degradation can also occur in certain medical conditions (multiple sclerosis being the prime example).

But how quickly the signals move is only part of the story — the other part is how far the signals have to travel. Axons can be very long, but information moves more quickly when the connection between two neurons is very short. Consider how many neurons you need to activate to have a coherent thought, and you'll realize that you'll do your best thinking when the neurons you need are all clustered tightly together.

Here's the last crucial concept: a neuron doesn't care what information it carries; a neuron, like your brain, is flexible. However, if you keep sending the same (or similar) information through, a small network of closely arranged neurons will develop to carry that specific information. With practice (the frequency effect), the connections between the neurons will grow 'stronger' — more used to carrying that information across particular synapses, more easily activated when triggered.

Working memory — a constraining factor

It seems incredible that we can store all the memories we accumulate in such a system, but we have some 200 billion neurons in our brain, and each neuron has about 1000 synapses on average. These are unimaginable numbers. But although our memory store is vast, as in a real jungle we can't see very much of it at a time. In fact, it's quite remarkable how little we can 'see' at any point, and this limitation is one of the critical constraints on our learning and our understanding.

We call the tiny part of memory that we are aware of, **working memory**. When you put information into your memory, the encoding takes place in working memory. When you drag it out of your memory, you pull it into working memory. When you read, you are using working memory to hold each word long enough to understand the complete sentence. When you think, it is working memory that holds the thoughts you are thinking.

As we all know to our cost, working memory is very small. Try and hold an unfamiliar phone number in your mind long enough to dial it and you quickly realize this. Probably the most widely known 'fact' about working memory is that it can only hold around seven **chunks** of information (between 5 and 9, depending on the individual). But we know now that working memory is even smaller than that. The 'magic number 7' (as it has been called) applies to how much you can hold if you actively maintain it — that is, repeat it to yourself. In the absence of this deliberate circulation, it is now thought that working memory can only hold around four chunks (between three and five), of which only one can be attended to at any one time (that is, only one is 'in focus').

Although it sounds like a small difference, the difference between having a working memory capacity of three chunks or one that can hold five chunks has significant effects on your cognitive abilities. Your working memory capacity is closely related to what is called **fluid intelligence**, meaning the part of an IQ test that has nothing to do with knowledge but depends almost entirely on your ability to reason and think quickly.

While working memory capacity may seem to be a 'fixed' attribute, something you are born with, it does increase during childhood and adolescence, and tends to decrease in old age. There have been a number of attempts to increase people's working memory capacity through training, some of which have had a certain amount of success, but most of this success has been with people who have attention difficulties. It is much less clear that training can increase the working memory capacity of an individual without cognitive disabilities.

At a practical level, however, differences in working memory capacity have a lot to do with how well we form our memory codes — with our skill in leaving out irrelevant material, and our skill at binding together the important stuff into a tightly-bound network. This is implicit in the word 'chunks'.

Although working memory can hold only a very small number of chunks, 'chunks' is the escape clause, as it were — for what constitutes a chunk is a very flexible matter. For example, 1 2 3 4 5 6 7 are seven different chunks, if you remember each digit separately (as you would if you were not familiar with the digits, as a young child isn't). But for those of us who are well-versed in our numbers, 1 through to 7 could be a single chunk. Similarly, these nine words:

1. brown

2. the

3. jumps

4. dog

5. over

6. quick

7. lazy

8. the

9. fox

could be nine chunks, or, in a different order ("the quick brown fox jumps over the lazy dog"), one chunk (for those who know it well as an example used in typing practice). At a much higher level of expertise, a chess master may have whole complex sequences of chess moves as single chunks.

Think back to what I said about how clusters of neurons become more strongly and closely connected with practice, and you'll see why practice is the key to functionally increasing your working memory capacity. The key is your chunks. A chunk is a very tight cluster. Such clusters enable you to increase how much information you can 'hold' in working memory.

I said that working memory contains a certain number of chunks, but to a large extent this way of thinking about it is a matter of convenience. It's more precise to say that the amount of information you can hold depends on how fast you are moving it. Because here's the thing about working memory — nothing stays in it for more than a couple of seconds, if you're not consciously keeping it active. This is why you have to keep repeating that phone number: you have to bring each digit back into 'focus' before its time is up and it fades back into the long-term store.

Let's go back to my earlier statement that the 'magic number 7' has now been reduced to 4 in the absence of deliberate repetition. We can reconcile these two numbers through another concept: that working memory (the 'inner circle') is surrounded by an outer area, in which, say, 3 items that have recently been in working memory can hover for a while, ready to be pulled back in easily.

Let's see that at work in a simple equation. Say you were given this problem to solve:

(38 x 4)/3

How you solve this will depend on your mathematical expertise, but a common way would be to break it down to:

30 x 4 = 120

8 x 4 = 32

120 + 32 = 152

150/3 = 50

2/3 = $^2/_3$

50 + $^2/_3$ = 50 $^2/_3$

Now think of the working memory flow needed to achieve this. First, you separate 38 into 30 and 8, moving your focus from 38 to 30, between 30 and 4 as you produce a new number (120) which briefly becomes the focus as you update working memory (30 and 4 can be discarded, replaced by this new number). Now you pull the waiting 8 into focus, perform the updating operation on it (multiply by 4), and replace it with the new number, 32. Now you must pull 120 back into focus as you add it to 32 and update working memory again, replacing 32 and 120 with the new number 152. Now you need the "divide by 3" (I hope it's still waiting! It all depends how fast you've been.) And so on.

Let's think about what you have to hold in working memory while you perform this fairly simple calculation:

30 x 4 = 120 (while performing this operation you need to hold in mind that 30 is just part of 38 and that you'll need to multiply the 8 by 4 and then add the two sums together to get a new sum that will then need to be divided by 3)

8 x 4 = 32 (performing this operation while holding in mind:
the previous sum (120); the need to add the two sums
together; the later need to divide by 3)

120 + 32 = 152 (holding in mind the division by 3)

150/3 = 50 (holding in mind the 2 left over)

2/3 = $^2/_3$ (holding in mind the 50)

50 + $^2/_3$ = 50 $^2/_3$

Here's a visual of the first step (the numbers floating in the
space beyond the rings represent information in your long-term memory store).

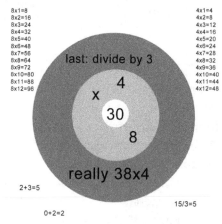

You can see working memory is already filling up, but if you're not skilled at math, each of these operations may involve a little more work, especially if you don't know your tables well, so that "32" doesn't come instantly to mind when you see "8 x 4". In such a case, there'll be even less space available for those other parts of the equation that you're holding in mind.

Moreover, if you suffer from 'math anxiety', you might also have other thoughts cluttering up the space — thoughts like "oh I'm hopeless at math", memories of failure, and the like.

On the other hand, you might have approached the equation this way:

38 doubled is 76 (at this point, you're holding in mind the
intention to double it again, and to divide the answer by 3)

76 doubled is 152 (at this point, you only need to hold in mind the intention to divide the answer by 3)

152/3 is 50 with 2 left over

50 $^2/_3$

Because working memory is fundamentally time-based, a lot comes down to how quickly you can perform basic operations — which is simply another way of saying how accessible the information is.

Because I had a teacher at school who used to put a number on the board and tell us to keep doubling (or halving) until he returned to the room, I have no trouble instantly transforming 38 to 76, 76 to 152. It would be a different story if I had to laboriously say to myself: 70 + 70 = 140, 6 + 6 = 12, 140 + 12 = 152.

This is why you can functionally (i.e., for practical purposes) increase your working memory capacity in specific areas by making your long-term knowledge more readily accessible — it all comes down your level of expertise, that is, practice.

Understanding that memory codes flow between long-term memory, the focus, the inner ring of working memory, and the outer ring of working memory, is critical for understanding the fundamental principles of effective practice. The flow is governed by time and attention. Once an item is out of focus, it only has a couple of seconds before it will retire out of the inner ring into the outer ring, unless you bring it back into focus. Once it's in the outer ring, it can only stay there a short time before it will fade back into the vast sea of long-term memory — unless, of course, you bring it back into focus.

It is by keeping the information moving, therefore, that you keep it in working memory. Fast speakers, and fast thinkers, have an advantage.

Understanding the limits of your working memory capacity helps you work out the most effective approach to your own practice, in particular, how much information you should try and handle at one time. It also helps you recognize when a particular learning task might be more demanding of your working memory.

The demands any cognitive task put on working memory we call **cognitive load**. Being able to assess the cognitive load of a task gives you the opportunity to reduce the load where necessary. Often, the reason a student has trouble understanding or performing a task is because the cognitive load is too much for them. Note that the cognitive load of a task is not a fixed amount, but varies for the individual, because, of course, it depends on the chunks (that is, on your long-term memory codes).

To a large extent, then, cognitive load is reflected in task difficulty.

Here are some factors to consider when trying to assess cognitive load:

- **how much information there is**

- **how complex the information is**
  How difficult it is to craft it into a tightly-bound cluster — that is, how hard it is for you to connect the information together and make sense of it.

- **how often you have to shift your focus**
  For example, if you are trying to do more than one task at a time (writing an essay and checking your Facebook page), then you are appreciably adding to your cognitive load; if you are trying to translate a text in a foreign language and have to keep diverting to the dictionary and a grammar book, then this will add

considerably to the cognitive load (as compared to being able to translate without needing to check vocabulary and grammar points).

- **whether information already in focus has to be altered and how much time and effort is needed to change it**
As in the math example, when you were performing calculations that 'updated' the number in focus.

There are basically two approaches to reducing cognitive load:

- break the task into smaller components
- practice the task or parts of the task until they are so easily retrieved from long-term memory that they are essentially automatic.

Thus, to reduce the cognitive load of dual essay-writing and Facebook-checking, you simply stop doing one activity and focus on the other. To reduce the cognitive load of trying to translate while constantly checking words and grammar, you need to practice your vocabulary and grammar until they are readily accessible.

Practice, then, is at the heart of reducing cognitive load and functionally improving your working memory capacity.

The role of consolidation in memory

But putting your code together in working memory is not all there is to encoding. Before your memory code can take its place in long-term memory, you need to **consolidate** it.

Memory consolidation is the final concept we need to understand how and why practice works.

Consolidation is aptly named. It's a process that takes place over time, and in a sense never truly ends. Your memories are always in a state of flux, vulnerable to change — every time you remember something, you change the memory code, even if only a little. But such changes can be thought of as *re*-consolidation. The initial stage in the consolidation process involves stabilizing the memory code, and this takes place over several hours. This is a crucial point when it comes to practicing effectively, because new memories are particularly vulnerable to being lost during this period.

The second stage of the memory-cycle is consolidation, and a lot of that occurs during sleep. During consolidation, the memory is edited — bits deemed unimportant may be dropped; important bits emphasized; connections made stronger or cut.

The time it takes to stabilize and consolidate new learning is why you often do better (perform a new skill better, or remember more) the next day.

encoding
initial process of transforming information into code

⬇

stabilization
< 6 hours

⬇

consolidation
mostly occurs during sleep

⬇

reconsolidation
happens every time you retrieve the memory code

It's also why it really helps if you run over any new learning in the evening before bed. If you learn in the morning, you have all your subsequent experiences and new information interfering with that earlier learning. New memory codes that haven't yet been consolidated are particularly vulnerable to interference, so they may well be over-written or corrupted by the time your sleeping brain looks over the day's events to see what is worthy of being kept.

Some types of information are more easily lost than others. For example, contextual details — such as knowing where you met a person; remembering the gender of the voice which spoke the foreign words you are trying to learn; remembering where a particular section is in a textbook — are particularly vulnerable. While such details might seem peripheral and not worth worrying about, context information is vital for source memory (remembering where and when you learned something), and also provides useful retrieval cues. So it's not to be despised, although you shouldn't waste too much attention on it either. The trick is to attend to useful context cues, and ignore less useful ones.

The third and final stage in the consolidation process — again, of particular interest in regard to effective practice — is re-consolidation, and that occurs every time the memory is retrieved from long-term memory. In other words, every time you practice or remember something, you are in effect re-making that memory code. This gives you a chance to improve it, to make it more memorable. Of course, it also means you can damage it — make it less what you want or need.

That's why practice has to be right. If you practice the wrong thing, the wrong information, then you make *that* more memorable, at the cost of the correct information.

This has been a very brief description of how memory works. I have provided only the most basic concepts that you need to know in order to understand how practice works and how to get the most out of it. If you want to know more about these principles, I discuss them in much more detail in my books *The Memory Key*, and its revised version *Perfect Memory Training*.

Points to remember

Memory is about two processes:

- **encoding** (shaping the memory and connecting it with existing memory codes) and

- **retrieving** (finding the memory by following the trail of connections).

A 'good' memory is one that is easily retrieved.

Repetition makes the memory trail stronger.

Recency *appears* to make the memory trail stronger, but this isn't a lasting strength.

Both encoding and retrieval involve working memory.

Working memory only holds a very small amount at a time. To keep a code there for more than 1.5-2 seconds, you need to keep shifting it into focus. This affects the content of your practice.

Your working memory capacity affects the cognitive load of a task, but you can reduce cognitive load by breaking down the task.

You can also reduce cognitive load by practicing the task (or parts of the task) until it becomes automatic, or almost so.

Memory codes aren't established in long-term memory until they're stabilized and consolidated.

Stabilization may take as long as six hours, and consolidation mainly occurs during sleep (daytime naps and even short periods of rest can also provide opportunities for some consolidation).

What should you practice?

> If you've ever seriously practiced a skill such as playing a sport or a musical instrument, you'll know very well that half the battle is pinpointing precisely *what* to practice. In this chapter, I discuss this issue in relation to two types of learning situation — the learning of relatively meaningless information (foreign language vocabulary) and the learning of reasonably complex text.

Before we get to the most effective ways of practicing, we need to look at the content of your practice. There is no point in diligently applying the principles of effective practice if you're practicing the wrong thing!

You may think it's self-evident, that of course you would practice what you need to practice. But unfortunately it is not always (or even often) that obvious.

Some examples

Let's start by looking at what would seem to be a very obvious situation. Say you want to learn ten new Spanish words:

el lápiz — pencil

la papelera — wastepaper bin

las tijeras — scissors

la regla — ruler

las cuentas — sums

el pupitre — desk

el techo — ceiling

la pared — wall

el suelo — floor

el pincel — paintbrush

The simplest method is to repeat them 'by rote': you say the pairs over and over again.

Bad idea. You waste a lot of time, and get little return.

Or you might be more cunning, tying the words together using a keyword-type mnemonic (if you're not familiar with this, don't worry, I discuss it in the next chapter). Thus, instead of repeating "las tijeras-scissors", you repeat "tiger scissors".

But here's the thing — what you need to practice is not the word-pairs, but the task you're aiming to master. You need to think about why you want to learn these words. What's your goal?

If you simply want to learn to read Spanish, your task is to be able to retrieve the English meaning when you see the Spanish word. So that's what your practice needs to be. You want to practice remembering "scissors" when you see "las tijeras". That is, you need to practice *retrieving* "scissors"; it is not enough to simply repeat the two together.

If you have the larger requirement of wanting to be able to communicate in Spanish — which means being able to read, write, speak, and understand Spanish when spoken —then you need to practice both directions. That is, you need to practice retrieving the English when you see the Spanish, and retrieving the Spanish when you see the English. So you need to practice recalling "scissors" when you see "las tijeras", and recalling "las tijeras" when you see "scissors". Practicing this in both modalities — the written word and the spoken word — is also a good idea.

This is the principle: **effective practice matches the task you want to master**.

There's another principle that follows from that: the more specific the task, the easier it is to practice; **the more general the task, the more varied your practice will need to be**.

Thus, for example, if you are only ever going to come across the word *regla* in connection with *cuentas*, then it would be fine to only practice retrieving the word *ruler* as part of this ordered list. But of course that's not going to happen. So you need to practice retrieving your words in different orders / contexts. And that means in completely different contexts. This group of words all relate to the classroom, and it's fine (indeed a good idea) to begin with categories. But it's not a good idea to stick with them forever. You want to practice retrieval in all the contexts in which you are likely to encounter the words (or at least a wide diversity of contexts).

Procedural learning — that is, the learning of a skill or procedure — also provides a relatively straightforward task. But again, you need to think about the contexts. Let's take the example of driving a car. Here in New Zealand, we recently changed a road rule governing who gives way to whom. It's a fairly major step changing such an important road rule and widespread chaos was predicted. It didn't, thankfully, eventuate

— the change was well-communicated to the public, and perhaps more importantly, it was an improvement on the old road rule. But I noticed one interesting difficulty. Although it was no great trouble to remember the new rule in relatively unfamiliar intersections, it took quite an effort of concentration to remember it at very familiar intersections. The problem is that we develop specific patterns of behavior in response to well-travelled situations, and these are only loosely associated with the more general, abstracted pattern, the 'rule'.

In other words, I needed to practice not only the general principle, but also the specific change in behavior at particular intersections ("when I'm *here* and wanting to turn *there*, I need to wait for any cars coming from *that direction*").

Of course, both vocabulary learning and skill learning provide reasonably cut-and-dried examples of the content you need to practice. The more common situation in academic study is that you have texts of varying complexity and you have to somehow work out what the information is that you need to learn. Consider, for example, this text on ozone and UV radiation:

The Relationship Of Ozone And Ultraviolet Radiation: Why Is Ozone So Important?

In this section, we will explore what is ozone and what is ultraviolet radiation. We then will explore the relationship between ozone and ultraviolet radiation from the sun. It is here that ozone plays its essential role in shielding the surface from harmful ultraviolet radiation. By screening out genetically destructive ultraviolet radiation from the Sun, ozone protects life on the surface of Earth. It is for this reason that ozone acquires an enormous importance. It is why we study it so extensively.

2.1 Ozone and the Ozone Layer

About 90% of the ozone in our atmosphere is contained

in the stratosphere, the region from about 10 to 50-km (32,000 to 164,000 feet) above Earth's surface. Ten percent of the ozone is contained in the troposphere, the lowest part of our atmosphere where all of our weather takes place. Measurements taken from instruments on the ground, flown on balloons, and operating in space show that ozone concentrations are greatest between about 15 and 30 km.

Although ozone concentrations are very small, typically only a few molecules O3 per million molecules of air, these ozone molecules are vitally important to life because they absorb the biologically harmful ultraviolet radiation from the Sun. There are three different types of ultraviolet (UV) radiation, based on the wavelength of the radiation. These are referred to as UV-a, UV-b, and UV-c. UV-c (red) is entirely screened out by ozone around 35 km altitude, while most UV-a (blue) reaches the surface, but it is not as genetically damaging, so we don't worry about it too much. It is the UV-b (green) radiation that can cause sunburn and that can also cause genetic damage, resulting in things like skin cancer, if exposure to it is prolonged. Ozone screens out most UV-b, but some reaches the surface. Were the ozone layer to decrease, more UV-b radiation would reach the surface, causing increased genetic damage to living things.

Because most of the ozone in our atmosphere is contained in the stratosphere, we refer to this region as the stratospheric ozone layer. In contrast to beneficial stratospheric ozone, tropospheric ozone is a pollutant found in high concentrations in smog. Though it too absorbs UV radiation, breathing it in high levels is unhealthy, even toxic. The high reactivity of ozone results in damage to the living tissue of plants and animals. This damage by heavy tropospheric ozone pollution is often manifested as eye and lung irritation. Tropospheric ozone is mainly produced during the daytime in polluted regions such as urban areas. Significant government efforts are underway to regulate the gases and emissions that lead to this harmful pollution, and smog alerts are regular occurrences in polluted urban areas.

2.2 Solar Radiation

To appreciate the importance of stratospheric ozone, we need to understand something of the Sun's output and how it impacts living systems. The Sun produces radiation at many different wavelengths. These are part of what is known as the electromagnetic (EM) spectrum. EM radiation includes everything from radio waves (very long wavelengths) to X-rays and gamma rays (very tiny wavelengths). EM radiation is classified by wavelength, which is a measure of how energetic is the radiation. The energy of a tiny piece or 'packet' of radiation (which we call a photon) is inversely proportional to its wavelength.

The human eye can detect wavelengths in the region of the spectrum from about 400 nm (nanometers or billionths of a meter) to about 700 nm. Not surprisingly, this is called the visible region of the spectrum. All the colors of light (red, orange, yellow, green, blue, and violet) fall inside a small wavelength band. Whereas radio waves have wavelengths on the order of meters, visible light waves have wavelengths on the order of billionths of a meter. Such a tiny unit is called a nanometer (1 nm= 10^{-9} m). At one end of the visible 'color' spectrum is red light. Red light has a wavelength of about 630 nm. Near the opposite end of the color spectrum is blue light, and at the very opposite end is violet light. Blue light has a wavelength of about 430 nm. Violet light has a wavelength of about 410 nm. Therefore, blue light is more energetic than red light because of its shorter wavelength, but it is less energetic than violet light, which has an even shorter wavelength. Radiation with wavelengths shorter than those of violet light is called ultraviolet radiation.

The Sun produces radiation that is mainly in the visible part of the electromagnetic spectrum. However, the Sun also generates radiation in ultraviolet (UV) part of the spectrum. UV wavelengths range from 1 to 400 nm. We are concerned about ultraviolet radiation because these rays are energetic enough to break the bonds of DNA molecules (the molecular carriers of our genetic coding), and thereby damage cells. While most plants and animals are able to either repair or destroy damaged cells, on occasion, these damaged DNA molecules are not repaired, and can replicate, leading to dangerous forms of skin cancer (basal, squamous, and melanoma).

2.3 Solar Fluxes

Solar flux refers to the amount of solar energy in watts falling perpendicularly on a surface one square centimeter, and the units are watts per cm^2 per nm. Because of the strong absorption of UV radiation by ozone in the stratosphere, the intensity decreases at lower altitudes in the atmosphere. In addition, while the energy of an individual photon is greater if it has a shorter wavelength, there are fewer photons at the shorter wavelengths, so the Sun's total energy output is less at the shorter wavelengths. Because of ozone, it is virtually impossible for solar ultraviolet to penetrate to Earth's surface. For radiation with a wavelength of 290 nm, the intensity at Earth's surface is 350 million times weaker than at the top of the atmosphere. If our eyes detected light at less than 290 nm instead of in the visible range, the world would be very dark because of the ozone absorption!

2.4 UV Radiation and the Screening Action by Ozone

To appreciate how important this ultraviolet radiation screening is, we can consider a characteristic of radiation damage called an action spectrum. An action spectrum gives us a measure of the relative effectiveness of radiation in generating a certain biological response over a range of wavelengths. This response might be erythema (sunburn), changes in plant growth, or changes in molecular DNA. Fortunately, where DNA is easily damaged (where there is a high probability), ozone strongly absorbs UV. At the longer wavelengths where ozone absorbs weakly, DNA damage is less likely. If there was a 10% decrease in ozone, the amount of DNA damaging UV would increase by about 22%. Considering that DNA damage can lead to maladies like skin cancer, it is clear that this absorption of the Sun's ultraviolet radiation by ozone is critical for our well-being.

While most of the ultraviolet radiation is absorbed by ozone, some does make it to Earth's surface. Typically, we classify ultraviolet radiation into three parts, UV-a (320-400 nm), UV-b (280-320 nm), and UV-c (200-280 nm). Sunscreens have been developed by commercial manufacturers to protect human skin from UV radiation. The labels of these sunscreens usually note that they screen both UV-a and UV-b. Why not also screen for UV-c radiation? When UV-c encounters ozone in the mid-stratosphere, it is quickly absorbed so that none reaches Earth's surface. UV-b is partially absorbed and UV-a is barely absorbed by ozone. Ozone is so effective at absorbing the extremely harmful UV-c that sunscreen manufacturers don't need to worry about UV-c. Manufacturers only need to eliminate skin absorption of damaging UV-b and less damaging UV-a radiation.

The screening of ultraviolet radiation by ozone depends on other factors, such as time of day and season. The angle of the Sun in the sky has a large effect on the UV radiation. When the Sun is directly overhead, the UV radiation comes straight down through our atmosphere and is only absorbed by overhead ozone. When the Sun is just slightly above the horizon at dawn and dusk, the UV radiation must pass through the atmosphere at an angle. Because the UV passes through a longer distance in the atmosphere, it encounters more ozone molecules and there is greater absorption and, consequently, less UV radiation striking the surface.

[adapted from NASA's Stratospheric Ozone Electronic Textbook, http://www.ccpo.odu.edu/SEES/ozone/oz_class.htm]

What should you do, faced with this dense text? How can you turn it into an information set that is 'learnable'?

One tried-and-true piece of advice is to find the 'main idea' in each paragraph. But this is rarely as straightforward as it sounds, and many students find this too difficult. They might instead decide to take the first (or last) sentence of each paragraph. Let's look at the 'learnable points' (that is, a set of points that you can use for revision) that would be generated from such a strategy.

first-sentence summary	last-sentence summary
90% of the ozone is in the stratosphere	ozone concentrations are greatest between about 15 and 30 km [presumably above the Earth's surface]
the stratosphere is the region from about 10 to 50-km (32,000 to 164,000 feet) above Earth's surface	reduction of ozone layer would increase UV-b radiation reaching the surface
ozone concentrations are very small	UV-b radiation causes genetic damage to living things

first-sentence summary	last-sentence summary
ozone is vital because it absorbs harmful ultraviolet radiation from the Sun	The energy of a tiny piece or 'packet' of radiation (which we call a photon) is inversely proportional to its wavelength. — definition
human eyes can see only part of the spectrum: from about 400 nm to 700	Radiation with wavelengths shorter than those of violet light is called ultraviolet radiation. — definition
Sun's radiation is mainly visible to us	the damage to DNA can usually be fixed, but when it isn't it can lead to skin cancer
Solar flux refers to the amount of solar energy in watts falling perpendicularly on a surface one square centimeter, and the units are watts per cm2 per nm. — definition	only UV-b, and to a lesser extent UV-a, matter
something called an action spectrum is a characteristic of radiation damage	
most but not all UV radiation is absorbed by ozone	
ozone's effectiveness depends on time of day and season	

There are 10 paragraphs (not counting the introductory one) in the original text. A 'main idea' summary would therefore be presumed to produce 10 learnable points, which indeed is what the first-sentence summary produces. The last-sentence summary, on the other hand, only produces 7 learnable points, because the last sentences in three of the paragraphs produced no learnable point (that is, they contained no important information). This suggests that the first-sentence summary is 'better', but think about this: if both sets of learnable points provide important information, then if you only have one point from each paragraph, you must be missing some important information.

So let's try instead to pick out the most important sentences regardless of where they are in the text, and not restricting ourselves to only one in each paragraph:

- By screening out genetically destructive ultraviolet radiation from the Sun, ozone protects life on the surface of Earth.

- About 90% of the ozone in our atmosphere is contained in the stratosphere, the region from about 10 to 50-km (32,000 to 164,000 feet) above Earth's surface.

- Ten percent of the ozone is contained in the troposphere, the lowest part of our atmosphere

- ozone concentrations are greatest between about 15 and 30 km.

- ozone molecules are vitally important to life because they absorb the biologically harmful ultraviolet radiation from the Sun. There are three different types of ultraviolet (UV) radiation, based on the wavelength of the radiation. These are called UV-a, UV-b, and UV-c.

- It is the UV-b radiation that can cause sunburn and that can also cause genetic damage,

- In contrast to beneficial stratospheric ozone, tropospheric ozone is a pollutant found in high concentrations in smog.

- The high reactivity of ozone results in damage to the living tissue of plants and animals.

- The Sun produces radiation at many different wavelengths.

- EM radiation is classified by wavelength, which is a measure of how energetic is the radiation.

- The energy of a tiny piece or 'packet' of radiation (which we call a photon) is inversely proportional to its wavelength.

- The human eye can detect wavelengths in the region of the spectrum from about 400 nm (nanometers or billionths of a meter) to about 700 nm.

- Violet light has a wavelength of about 410 nm.

- Radiation with wavelengths shorter than those of violet light is called ultraviolet radiation.

- We are concerned about ultraviolet radiation because these rays are energetic enough to break the bonds of DNA molecules (the molecular carriers of our genetic coding), and thereby damage cells.

- Solar flux refers to the amount of solar energy in watts falling perpendicularly on a surface one square centimeter, and the units are watts per cm^2 per nm.

- An action spectrum gives us a measure of the relative

effectiveness of radiation in generating a certain biological response over a range of wavelengths.

- If there was a 10% decrease in ozone, the amount of DNA damaging UV would increase by about 22%.

- we classify ultraviolet radiation into three parts, UV-a (320-400 nm), UV-b (280-320 nm), and UV-c (200-280 nm).

- When UV-c encounters ozone in the mid-stratosphere, it is quickly absorbed so that none reaches Earth's surface. UV-b is partially absorbed and UV-a is barely absorbed by ozone.

- The angle of the Sun in the sky has a large effect on the UV radiation.

Unfortunately, this still manages to occasionally miss out some important details that help us understand and remember. This is the problem with using verbatim text — the author doesn't always put the important information in the precise sentences you want; sometimes the information you want is spread across two or more sentences, but to include all the sentences that contain some useful nugget, you would end up with a text that's not that much shorter than your original text! No, your aim should be to get *all* the vital information, but *only* the vital information. To do that, you need to paraphrase.

Here's my paraphrased summary:

Ozone is important because it shields the surface from harmful ultraviolet radiation.

About 90% of the ozone in our atmosphere is contained in the stratosphere (the ozone layer), and 10% in the troposphere, the lowest part of our atmosphere where all of our weather takes place.

There are three different types of ultraviolet (UV) radiation, based on the wavelength of the radiation: UV-a (longest), UV-b, and UV-c (shortest). UV-c is entirely screened out by the ozone layer, and UV-a is not as damaging, so the main problem is UV-b.

Tropospheric ozone is a pollutant found in high concentrations in smog. The high reactivity of ozone results in damage to the living tissue of plants and animals, and is often felt as eye and lung irritation.

The Sun produces radiation at many different wavelengths. Electromagnetic radiation is classified by wavelength, which is a measure of how energetic is the radiation.

The visible part of the electromagnetic spectrum ranges from 400 nanometers to 700 nm. Red light has a wavelength of about 630 nm; violet light about 410 nm. Radiation with wavelengths shorter than those of violet light is called ultraviolet radiation.

Ultraviolet rays are energetic enough to break the bonds of DNA molecules, and thereby damage cells. While our bodies can repair this most of the time, sometimes damaged DNA molecules are not repaired, and can replicate, leading to skin cancer.

Solar flux refers to the amount of solar energy in watts falling perpendicularly on a surface one square centimeter, and the units are watts per cm^2 per nm. The strong absorption of UV radiation in the ozone layer reduces the intensity of solar energy at lower altitudes. More energetic photons (ones with shorter wavelengths) are also less common.

The action spectrum measures the relative effectiveness of radiation in generating a certain biological response (such as sunburn) over a range of wavelengths. Because ozone is most protective on the most dangerous wavelengths, a 10% decrease in ozone would increase the amount of DNA-damaging UV by about 22%.

Time and season affect how much UV radiation is absorbed by ozone because the angle of the sun affects how long the radiation takes to pass through the atmosphere (the path is shorter when the sun is directly overhead, so the radiation meets fewer ozone molecules).

This is still quite lengthy (380 words compared to our original text of 1400 words), but now that we have all the important information, we can turn it into a good learnable set. To do that, you need to reduce the summary further, and the more you already know about the topic, the more you can reduce it. Here's my learnable set, assuming a little knowledge of electromagnetic radiation:

- Ozone is important because it shields the surface from harmful ultraviolet radiation.

- The stratosphere holds 90% of the ozone in our atmosphere (the ozone layer).

- The troposphere holds 10%.

- The ozone layer protects us; tropospheric ozone is a pollutant found in high concentrations in smog.

- Radiation with wavelengths shorter than those of violet light (at the short end of the visible spectrum) is called ultraviolet radiation. UV waves are dangerous because they're energetic enough to break the bonds of DNA molecules.

- Of the three different types of ultraviolet (UV) radiation, the shortest (UV-c) is entirely screened out by the ozone layer, while the longest (UV-a) is not so damaging, so the main problem is UV-b.

- Because ozone is most protective on the most dangerous wavelengths, a 10% decrease in ozone would increase the amount of DNA-damaging UV by about 22%.

- Time and season affect how much UV radiation is absorbed by ozone because the angle of the sun affects how long the radiation takes to pass through the atmosphere (the path is shorter when the sun is directly overhead, so the radiation meets fewer ozone molecules).

- Measurement:
  - Solar flux = the amount of solar energy in watts falling perpendicularly on a surface one square centimeter.
  - The action spectrum measures the relative effectiveness of radiation in generating a certain biological response (such as sunburn) over a range of wavelengths.

We've now reduced our original text to a mere 240 words, and the learnable set can easily be transformed into a Q & A format, perfect for practice:

Q: Why is ozone important?

A: Because it shields the surface from harmful ultraviolet radiation.

Q: What proportion of the atmosphere's ozone is in the stratosphere?

A: 90%

Q: In the troposphere?

A: 10%

Q: Is ozone always protective?

A: No. The ozone layer protects us, but tropospheric ozone is a pollutant found in high concentrations in smog.

Q: What is ultraviolet radiation?

A: Radiation with wavelengths shorter than those of violet light (at the short end of the visible spectrum).

Q: Why is it dangerous?

A: Because it's energetic enough to break the bonds of DNA molecules.

Q: Which of the three different types of ultraviolet radiation is most dangerous and why?

A: UV-b. Because the shortest (UV-c) is entirely screened out by the ozone layer, and the longest (UV-a) is not so damaging.

Q: How much would a 10% decrease in ozone increase the amount of DNA-damaging UV?

A: By about 22%.

Q: Why does time of day and season affect how much UV radiation is absorbed by ozone?

A: Because the angle of the sun affects how long the radiation takes to pass through the atmosphere (the path is shorter when the sun is directly overhead, so the radiation meets fewer ozone molecules).

Q: What is solar flux?

A: The amount of solar energy in watts falling perpendicularly on a surface one square centimeter.

Q: What does the action spectrum measure?

A: The relative effectiveness of radiation in generating a

certain biological response (such as sunburn) over a range of wavelengths.

Now, your learnable points may not be the same as the ones I've produced, because what's appropriate will depend on both your existing knowledge and your goals (affected by the extent of your interest in this topic, and the relevance of it to your other work and interests). But the point of this exercise is not to provide the 'correct' set of important points, but to demonstrate that the first and most important principle of effective practice is to work out exactly *what* you should be practicing, and to show that this isn't always intuitively obvious.

If you want more help with this very difficult skill, my books on note-taking cover this situation in considerable detail.

Points to remember

Effective practice matches the task you want to master.

The more specific the task, the easier it is to practice; the more general the task, the more varied your practice will need to be.

For complex text, you need to produce a set of learnable points for revision. This involves a 3-step process:

1. Highlight important information in the text that is unknown to you.

2. Re-organize and reduce the information, using your own words, to create a summary that succinctly captures what is important to you.

3. Transform your learnable set into a Q & A format.

Retrieval practice

> The reason for including 'revision' under the term 'practice' now becomes obvious. This chapter introduces the concept of 'retrieval practice' — the single most useful learning strategy there is. In this chapter, I explain what retrieval practice is and how it compares with other learning strategies. I also discuss whether mistakes matter.

I said earlier that effective practice matches the task you want to master. In the most general sense, the essence of that task is usually that you want to retrieve information from your long-term memory (the exception is if you are only interested in recognizing the correct information — recognition is a different process from retrieval). For the most part, though, when you practice or revise something, what you need to be doing is practicing retrieving the information. This is, unsurprisingly, called **retrieval practice**. The vital thing to remember is that it is not the same as repetition or rehearsal. The idea is not to simply *repeat* the correct information, but to try and *retrieve* it.

Clearly retrieval practice is a form of testing, and the effectiveness of retrieval practice as a means of learning is the main reason why repeated testing is valuable (far more clearly valuable than repeated testing as a means of assessment).

The basic memory principles I have discussed makes it clear why retrieval practice is so powerful — every time you retrieve the information, you make the trail to it stronger. But you must truly be retrieving it from long-term memory — if the information is still in working memory, you are simply keeping it active (in the same way that repeating a phone number to yourself keeps it in working memory), not retrieving it.

Retrieval practice is the single most powerful learning strategy there is.

Comparison of retrieval practice with other strategies

You almost certainly find it hard to believe that retrieval practice is as effective as it is. Even I've found myself, despite knowing all the research, surprised by the actual experience of learning using this method, and only this method. In truth, you're only going to be totally convinced by using the strategy.

However, since a crucial factor in whether or not a person uses an effective memory strategy is the extent to which they are convinced it is effective, I'm going to tell you about some of the research comparing the effectiveness of retrieval practice against other study strategies.

This discussion will also give you some idea how the strategy works in practice.

Re-reading

The most important comparison is with re-reading, because (unfortunately) this is the most common study strategy used by students. In a study in which college students were tested on their recall of two short prose passages, each about 250-275 words long, the students studied the texts in one half-hour session. During each of four 7-minute periods, they either read a text, re-read one of the texts, or took a recall test on the text they didn't re-read — meaning that one passage was read twice, while the other was read once and tested once.

When tested a mere five minutes after the study session, the passage read twice was recalled slightly better (the recency effect in action). However, when tested a week later, the passage that had been read once and tested once (not counting the 5-minute-delay test, on which no feedback was given) was remembered decidedly better than the one that had been read twice. Those re-reading scored 81% on an immediate test, but only 42% a week later. Those who read it only once, followed by a test, scored only 75% immediately, but 56% a week later (the difference, I note, between a passing grade and a fail!).

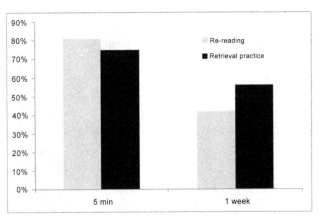

Figure 3.1: Mean proportion of idea units recalled 5 minutes after study and one week after study, for students who re-read compared to those who used retrieval practice. From Roediger & Karpicke, 2006.

In other words, while re-reading gave the immediate illusion of having been learned better, it was forgotten at a much greater rate over time (and bear in mind that this is only after a week;

the gap is expected to get wider over time).

In a further experiment, using only one of the two prose passages, some students read and re-read the passage during four 5-minute study periods, while another group studied their passage for three of the periods then were tested during the fourth, and the final group studied their passage during the first period before being given three recall tests.

Again, those students who only re-read the passage had an advantage when tested five minutes after the session. But, as before, the story was different a week later — and the best performers were those who had read the passage only *once*, followed by three tests. Those who only re-read the text scored 83% immediately, but only 40% after a week, while those who read the text only once, followed by three tests, scored 71% immediately but 61% a week later.

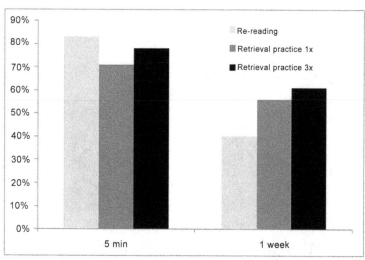

Figure 3.2: Mean proportion of idea units recalled 5 minutes after study and one week after study, for students who re-read compared to those who used retrieval practice once and those who used retrieval practice three times. From Roediger & Karpicke, 2006.

The benefit of retrieval practice is even more dramatic if you use a proportional measure in order to show how much *forgetting* took place. This shows 52% forgetting for the re-reading group compared to only 14% forgetting for the group who only read it once but were tested three times.

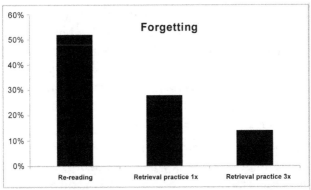

Figure 3.3: Amount of forgetting over one week, for students who re-read compared to those who used retrieval practice once and those who used retrieval practice three times. From Roediger & Karpicke, 2006.

Note that the re-reading group read the passage about 14 times in total, while the repeated testing (retrieval practice) group read the passage only an average of 3.4 times in its one-and-only study session. So, those re-reading experienced the material four times more than the testing group, but still failed to achieve much long-term remembering.

Despite this, those in the re-reading group were much more likely to believe they'd remember the information than students in the repeated-testing group. After all, the text is now very familiar, and when you have made no effort to test your ability to retrieve it, you have no idea how hard that might be.

Students re-read because they believe it is an effective strategy. They are wrong.

Let's look at some comparisons with more sophisticated and effective study strategies.

Keyword mnemonic

The keyword mnemonic is a much-studied and much-touted strategy for learning new words, most often in another language but not necessarily so. The essence of the keyword mnemonic (I discuss it later in more detail) lies in the choosing of an intermediary word that binds what you need to remember to something you already know well. So, to remember that the Spanish word *carta* means *letter* (the sort you post), you select an English word that sounds as close to *carta* as you can get, and you make a mental picture that links that word to the English meaning — for example, a letter in a cart.

Results from using the keyword method have been dramatic. For example, in the classic study, over a third of 120 Russian words were remembered more than 80% of the time in the keyword condition, compared to only one item in the control condition (*glaz* for *eye* — a mnemonic link so obvious I am sure most of the control participants used it even without explicit direction). Moreover, only seven words were remembered less than half the time in the keyword condition, compared to 70 in the control ("use your own method") condition!

Overall, the keyword group recalled 72% of the words when they were tested on the day following the three study days (40 words were studied each day), compared to 46% by the control group. When they were (without warning) tested again six weeks later, the keyword group remembered 43% compared to the control group's 28%. Compelling results.

Since that early study, there have been many experiments demonstrating the effectiveness of the keyword method, especially when measured against rote repetition or "use your own method", but also when compared with the popular context method (experiencing the word to be learned in several different meaningful contexts; guessing the meaning from the context).

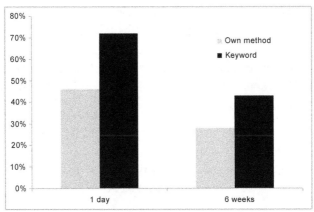

Figure 3.4: Average percentage of 120 Russian words recalled, one day after study and 6 weeks after study, for students using their own method to learn compared with those using the keyword mnemonic. From Atkinson 1975.

It is therefore very impressive that retrieval practice has been shown to be as effective, and sometimes even better, than the keyword method.

One study, for example, comparing the learning of 20 German words using either the keyword mnemonic, retrieval practice, or rote repetition, found equal levels of recall a day later for those using the keyword method and those using retrieval practice, with both significantly better than the group average for those using rote repetition (an average of 15 words vs 11).

A follow-up compared the learning of 24 German words in which sets of six words were learned in one of four ways:

- **Elaboration**: either describing a different English meaning for the word (e.g., "The German for SHARP is SCHARF, scharf also means hot (as in spicy).") or by breaking down a compound word into its components (e.g., "The German for LIGHTHOUSE is LEUCHTTURM, Leuchtturm consists of the two words for shine and tower.").

- **Retrieval practice**: filler pages between each retrieval attempt gave an expanding schedule of 1-3-5-7 (I will discuss practice schedules in a later chapter; what this means here is that there was one intervening filler item before the first retrieval attempt, three items before the second attempt, and so on).

- **Keyword mnemonic**: the English and German words were presented with a description of a suggested image (e.g., "The German for SHARP is SCHARF. Imagine cutting a German flag with SHARP scissors.").

- **Retrieval practice + keyword mnemonic**

As you can see from the following graphs, the elaboration strategy produced significantly worse learning than the others, and this was especially true with the more difficult task of remembering the German on seeing the English. On the easier task of recognizing the meaning of a German word, there was little difference between the other strategies on the immediate test, but when tested a week later, the combined strategy was significantly better than the others, and retrieval practice produced slightly, but not quite significantly, better recall than the keyword method. (The main problem with this study is that, for practical reasons, the number of words learned by each strategy is so small that it's hard to get a lot of difference between the recall scores.)

For the harder task (translating into German), retrieval practice achieved significantly better remembering than the keyword mnemonic.

In another study, students learned 48 Swahili words by the keyword method, with some students simply 're-studying' the words on practice trials (i.e., keyword method alone), while others practiced retrieving them (keyword + retrieval practice).

Those who were given retrieval practice performed almost *three times* better on the final test compared to those given restudy only: 40% correct vs 14%.

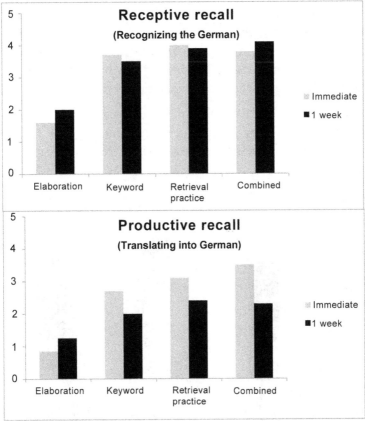

Figure 3.5: Average number of correct words out of six, for students using different learning strategies. From Fritz et al. 2007.

A triumph for the use of retrieval practice! But the real interest of this study lies in a further comparison they made. On the final test one week later, students were either given the cue only (the Swahili word), or the cue plus keyword, or the cue plus a prompt to remember their keyword.

The group that used testing as part of their study weren't significantly affected by the cue given in the final test — it didn't matter that much whether it was only the word itself, the word plus keyword, or the word with a reminder to help them recall their keyword. But the group that only restudied the material were significantly helped by being given the keyword as well as the cue. You can see in the graph below how badly the study-only group did when given only the cue, and how much it improved their performance when reminded of the keyword as well.

Moreover, when the researchers looked deeper into the results for the group receiving the prompt, they found that remembering the keyword made a huge difference to recall of the English word:

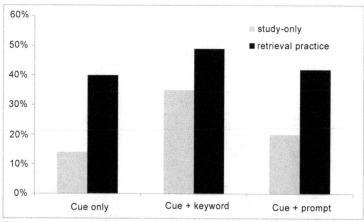

Figure 3.6: Average percentage of items correctly recalled on final test, when given either the cue only, the cue plus the keyword, or the cue plus a prompt to remember the keyword, for students who restudied only compared with those who used retrieval practice. From Pyc & Rawson. 2010.

You can see how recall of the English word depends very heavily on remembering the keyword. But the main point of this graph is the difference between the retrieval practice group and the study-only group when the keyword was remembered: see how the keyword triggered recall of the English word nearly 70% of the time for the retrieval practice group, but only around a third of the time for the study-only group. Why were the keywords so much more effective for the retrieval practice group? It seems likely that they had better keywords — keywords that did a better job of evoking the English words.

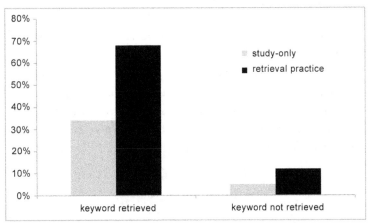

Figure 3.7: Average percentage of items correctly recalled on final test by students given the cue plus a prompt to remember the keyword, as a function of whether or not the keywords were remembered, for students who restudied only compared with those who used retrieval practice. From Pyc & Rawson. 2010.

In other words, retrieval practice supports the keyword method not simply because the path to the keywords has become stronger, but also because failures during testing are likely to encourage you to create better keywords. When you practice retrieval, you can see whether the keyword is a good one for you; if you're having trouble remembering the keyword, you know to try and find a better one.

This advantage of retrieval practice extends beyond the keyword method, of course. When you test yourself, you give yourself the opportunity to see how hard the material is to remember. This shows you when you need to augment retrieval practice with other methods, such as the keyword method or concept maps (below). Or, if you're already using such strategies, when you need to tweak (the keyword; the map) to make it more memorable.

Concept maps

Another very effective learning strategy, for a different type of learning situation, involves constructing concept maps. A concept map is a diagram that attempts to show how facts and ideas are connected to each other, through the use of labeled nodes (representing concepts) and lines connecting them (showing the relationships between the concepts). You are no doubt familiar with one type of concept map — the mind map.

Here's an example of a concept map (taken from *Effective notetaking*).

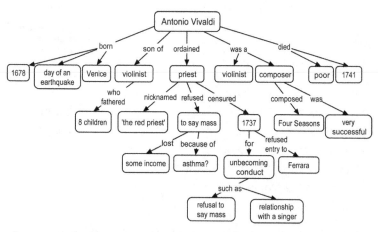

In one study comparing concept maps and retrieval practice, students studied a short text about sea otters using one of four strategies:

- **study-only**: this group studied the text for five minutes

- **repeated study**: this group studied the text during four 5-minute periods, each period separated by a one-minute break

- **concept mapping**: this group studied the text for five minutes, then were given 25 minutes to construct a concept map from the text, which they had in front of them

- **retrieval practice**: this group were given five minutes to study the text, then given 10 minutes to write down as much as they could recall; they were then given another five minutes to re-study the text, before carrying out a second recall test.

When tested a week later, performance was (as expected) worst in the study-once group, and best in the retrieval practice group — about 50% better than that of the concept mapping group (67% vs 45%).

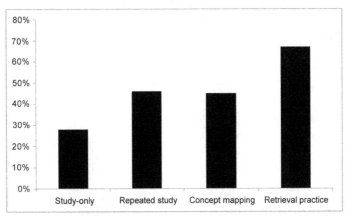

Figure 3.8: Average percentage of correctly answered short-answer questions after studying a science text either over one study period, or on four consecutive periods, or by creating a concept map or using retrieval practice after the initial study period. From Karpicke & Blunt. 2011.

It's interesting that there was no significant difference between the repeated-study and the concept mapping groups. This suggests that the main benefit of concept mapping (when used in this way, in the presence of the text, which is a very important point that I'll discuss later) is simply to increase the time spent with the material. Nothing wrong with that. Drawing a concept map is probably a more interesting way of spending time with the text than simply re-reading it, and it's always worth finding more interesting ways to do things (it increases the likelihood of you doing them!).

As always, when asked, students' beliefs were at odds with the results. Students in the repeated study condition gave the highest predictions for their learning, and those in the retrieval practice condition gave the lowest.

In another experiment, with students employing a concept map strategy on one text and retrieval practice on a different text, retrieval practice once again greatly out-performed concept mapping on the final short-answer test (an average score of 73% compared to an average of 54%) — and, surprisingly, also out-performed it on an alternative final test that involved drawing a concept map (44% vs 28%).

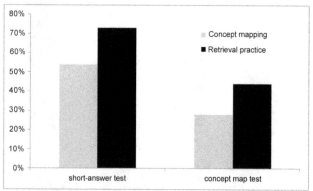

Figure 3.9: Average percentage correct on the short-answer questions and the concept map test, for students who had either created a concept map or used retrieval practice after the initial study period. From Karpicke & Blunt. 2011.

The study has one other important finding, concerning how the strategies suited individual students. While the great majority of students (101 of the 120; 84%) performed better after retrieval practice, a few (16%) benefited more from concept mapping. This reminds us that however effective a strategy is in general, it's not necessarily the best strategy for everyone.

I would also like to emphasize that these results should certainly not be taken as a slur on concept mapping, just as the earlier findings I discussed shouldn't be taken as an indictment of the keyword mnemonic. The point is rather that retrieval practice is a strategy that is of similar effectiveness, and greater effectiveness in some circumstances. Being an effective student, I must emphasize, is not about having one really effective study strategy, but about having a tool-box of effective strategies that you can choose from depending on the situation.

Having said that, if you only had one memory tool, retrieval practice would be the one to have!

The big advantage of retrieval practice is of course that it is a very simple, easily learned technique. It also requires much less cognitive effort than, say, the keyword mnemonic, which puts many people off because of the difficulty of finding good keywords, and the effort (which is greater for some than for others) of creating images.

But, as I say, this is not an either/or situation. In the case of learning new vocabulary, for example, the most effective strategy is to combine retrieval practice and the keyword mnemonic, using the principles of effective retrieval practice (which I will discuss in the next section) in conjunction with the keyword mnemonic, as needed. By 'as needed', I mean that you shouldn't feel the need to use the keyword mnemonic with every word. The effectiveness of retrieval practice on its own is sufficient for many words, and the effort of creating keywords is too great for many people to bother with using the mnemonic unless the word is of proven difficulty to remember. I will discuss this in more detail later.

Similarly, concept mapping can also be used very effectively as a retrieval practice technique. Again, I will discuss this in a later section.

Benefits for related information

Basically, retrieval practice is about testing yourself. There are two intriguing and unexpected aspects to this testing. One is that it can be of some benefit even when you get no feedback about whether your answer is right or wrong (but of course it's much more effective with feedback). Another is that it can improve your memory for untested material (but not just any untested information!).

This has been shown in a series of three experiments using much longer texts than typically used in lab experiments. The first experiment used a text of 2,700 words, and students either read the article then were tested on it, or read the article and then read the answers to the test questions. Those who were tested were given instructions not to guess, and were given no feedback on the accuracy of their answers. After they'd been tested, they were given the test again, with the questions in a different order. All the students were tested a day later.

Here's the crucial point — each question in the practice test was matched with a related question. For example, the question "Where do toucans sleep at night?" was related to "What other bird species is the toucan related to?" because toucans sleep in the holes created by woodpeckers. While most related pairs appeared in the same paragraph of the article, knowing the answer to one question wouldn't of itself tell you the answer to the other.

So what happened? Those given extra study recalled more of the practice-test questions (79% vs 70%) than those who were tested (remember that those in the testing condition weren't given any feedback on their answers, while those given extra study read the answers to the practice-test questions, so those given extra study had this advantage over the study-test group). However, and this is the interesting bit, the study-test group did better on the related questions (59% vs 49%) — that is, those that hadn't been part of the practice test. Similar findings occurred in the follow-up experiments.

Now I must emphasize that the information actually practiced is, of course, remembered far better than any other. But what this study tells us is that retrieval practice can have benefits that extend beyond the exact information practiced, to information that is related or appears in close proximity to the information practiced.

However, as the study also reveals (in its third experiment), this benefit applied only when the students searched their minds for related information when retrieving. If you keep a narrow focus on the exact question, retrieval benefits tend to be restricted just to the information retrieved. This is also reflected in time taken — students who were faster in retrieving the information during practice tended not to gain related benefits.

Errorless learning

This benefit of retrieval practice for related, untested information is called **retrieval-induced facilitation**. There's an inverse: **retrieval-induced forgetting**. This is reflected in an idea that's become popular in education — the idea that if you make a mistake when answering a question or solving a problem, then you make it harder for yourself to remember the correct information.

This idea has produced the idea of 'errorless learning', of thinking you need to ensure that learning situations provide the correct answers in order to prevent students producing a wrong answer. Errorless learning was born in animal training studies, and certainly it has proven helpful in rehabilitation for people with brain damage. However, it is less clear how far those findings can be extended to normal learning in humans.

The theoretical basis of errorless learning is reasonable enough — most of our memory failures do stem from interference from competing memories, so it's not unreasonable to expect that any mistakes you make will be remembered and will then compete with the correct answer. Indeed, it's undeniable that this does happen.

One example is with tip-of-the-tongue (TOT) problems (when you can't immediately retrieve a word, but feel it's 'on the tip of my tongue'). When experiencing this situation, we often produce incorrect words, which then can block retrieval of the correct word ("I know I know the name of the singer, but all I can think of is Joni Mitchell!"). Sometimes, when we produce the wrong words, we make it more likely that the next time we try and think of the word, we'll once again come up with the same incorrect one.

Remember that memory codes are reconsolidated every time you retrieve them. This provides an opportunity for any mistake made when trying to retrieve a word to attach itself to the memory code, or the path to the code. In other words, when you make a mistake, you risk learning that mistake. Research indicates that this is more likely to occur the more time you spend trying to retrieve the word.

But if retrieving the wrong information hurts our learning of the correct information, why is retrieval practice so effective? Doesn't retrieval practice set you up for occasional failures?

Recent research has shown that retrieval-induced forgetting only occurs within a brief time-frame. As long as there's a day's delay between the retrieval practice and the final test, it isn't an issue. (The delay necessary is probably less than that, but we have to go with the comparisons investigated. We know such forgetting can be observed when the test is up to 20 minutes after the retrieval practice; we know it isn't observed when there's a 24-hour delay.)

Additionally, retrieval-induced forgetting doesn't occur if you make multiple connections between list items (which is what naturally occurs when you're working through a text — you make multiple connections between the facts and ideas in the text). It's also reduced if you engage in distinctive processing for each word (such as thinking about how it's different from the other words).

In other words, if you integrate your information (as you do when reading a text), or if some time elapses between your retrieval practice and testing, then retrieval-induced forgetting should not be an issue. These (rather than the quick word-pair learning tasks so beloved in the laboratory) are rather more usual learning situations!

Moreover, even in the case where you are learning something like word pairs (such as when practicing foreign vocabulary), retrieval-induced forgetting only seems to occur when items are presented in a random order during retrieval practice. If the same order is maintained during practice as occurred during the original presentation (even if there are gaps from omitted items), then the effect doesn't occur.

In other words, retrieval-induced forgetting, which is used to justify an errorless learning strategy, only appears to occur in a narrow set of circumstances:

- when your learning — retrieval practice and testing — occurs within a very short time period

- when you don't make any attempts to understand or elaborate the material to be learned

- when you practice by mixing up the order of your items (note that this is generally a good idea, just not when the first two factors are in play).

There's also another factor: feedback.

Forced guessing

One question students and teachers often wonder about is whether it's a bad idea to force a guess (that is, even when you have absolutely no idea of the correct answer). A study that explored this question had students read 80 obscure facts before being tested on them. If they offered an answer, they were asked to rate their confidence in it. If they didn't offer an answer, they were randomly asked half the time to guess. They were told the correct answer in all cases, and were tested again a day later.

In almost all cases, questions answered correctly on the first test were again answered correctly (91%). Only 36% of those initially answered incorrectly were answered correctly on the final test. Only 19% of those questions in which the students were forced to guess, were answered correctly on the final test. This compares to 23% of those questions students were allowed to skip (a non-significant difference).

There are two particularly interesting findings:

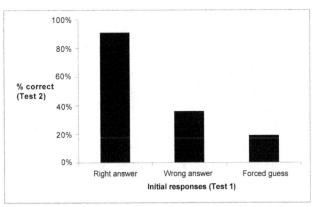

Figure 3.10: Average percentage correct on final test as a function of whether or not the student gave the right answer, the wrong answer, or was asked to guess, on the initial test. From Kang et al. 2011.

- Those who were more confident of their *incorrect* answers were *more* likely to get the answer correct on the final test.

- Forced guessing didn't make a significant difference, either way.

The relationship with confidence has been found before. It's even got a name. It's called the **hypercorrection effect**, and it's not as counter-intuitive as it may immediately appear. It may be, for example, that greater confidence produces greater surprise when the answer turns out to be wrong, and it is this surprise that makes you more likely to remember it.

On the other hand (and demonstrating that learning is a complex process, in which at any time a number of variables are potentially at work), incorrect answers at the *lowest* confidence level were also associated with better learning, while *correct* responses at the lowest confidence level (which might also be expected to elicit surprise) were *not* better learned.

It seems likely that there is more than one thing going on here. One is surprise; the other might have to do with your familiarity with the general area of knowledge in question. Your knowledge of the topic often affects a strategy's relative effectiveness — this is one reason why we can't just make simple black-and-white rules about what strategies are best to use. In this case, those with more familiarity with the topic might be more willing to offer low-confidence guesses. If they have higher domain knowledge, they'll be better able to acquire new information in it (this is why experts can learn new information in their area of expertise so much more easily than novices — they have a larger and denser network into which new information can slot).

But the finding of particular interest in respect to errorless learning is that forcing a guess did *not* significantly impair learning.

This may have been because immediate feedback was offered. Accordingly, a second experiment was run in which feedback wasn't offered until all 80 questions had been answered. The result was the same: no significant effect of forced guessing.

A similar experiment using explanations that required longer answers (e.g., "Why does the moon influence the Earth's tides more than the sun, even though the sun has the greater gravitational pull?") also found no effects of forced guessing.

One potentially important factor in these experiments is that, if people are being forced to guess, they probably are well aware that their guesses are likely to be wrong. This is a very different situation from making errors that you genuinely believe are correct. (Note, for example, that in this study only 36% of questions answered wrongly on the first test were corrected on the second.)

Nevertheless, for our purposes here, one thing is clear: guessing, even though it produces incorrect information, doesn't significantly harm learning, and this remains true even when feedback is a little delayed.

Having said that, it does seem likely that motor skills are quite a different story. They are far more likely to be adversely affected by incorrect actions, given that physical actions are recorded 'in the body', quite differently from declarative (factual) memory.

It's also possible that learning tasks that rely heavily on familiarity (such as multi-choice tests that only require recognition) might also be more badly affected by errors.

Points to remember

Testing can involve recognition or retrieval. A multi-choice test only requires recognition of the correct answer. A short-answer test requires you to retrieve it.

Retrieval practice is a study strategy in which you practice retrieving the information from your memory.

In comparison with other study strategies, retrieval practice has consistently been shown to be more effective.

The most popular study strategy is re-reading, which produces much faster forgetting.

Concept mapping *when used in the presence of the text* appears to produce no better learning than re-reading.

The keyword method is a very effective strategy for learning vocabulary, but retrieval practice seems to be even better. Combining the two is probably best of all.

Individual differences always matter. Some individuals may find another study strategy works better for them — but would probably find combining their preferred method with retrieval practice even more effective.

Retrieval practice can also benefit information that is related or appears in close proximity to the information practiced — but only if you search your mind for related information when retrieving.

Making errors when retrieving *can* corrupt your memory code, but this is mainly restricted to physical skill learning and simple, relatively meaningless information.

How often should you practice? 4

In this chapter, I get down to some of the practical aspects of practicing effectively: in particular, the number of times you should practice retrieving an item, both within a learning session and across sessions.

First and foremost among the questions of how to practice effectively is this one: how often do I need to practice? There are two parts to the answer:

- how often you should practice during your first study session

- how many times you should give yourself practice tests after that.

Criterion levels set the number of correct retrievals

Let's start the discussion with a bit of jargon — the concept of 'criterion level'. In psychology experiments, this refers to a specific level of performance that the participant must reach before moving on. In learning experiments, it usually means that the participant must correctly retrieve the information a certain number of times before they are judged to have 'learned' it.

Do pay special attention to that phrase *'correctly* retrieve'! Previously, learning research has tended to vary the number of retrieval *attempts*. Similarly, many students have some rule, such as repeating an item three times. But the vital part of this concept is that *only the correct retrievals count.*

A couple of studies have played with different criterion levels in an attempt to work out the best number of correct retrievals for efficient learning.

Task difficulty affects optimal criterion level

In one study, students studied 50 English-Lithuanian word pairs, which were displayed on a screen one by one for 10 seconds. After studying the list, the students practiced retrieving the English words. They were given little time to ponder — they had a mere eight seconds to type in the English word as each Lithuanian word appeared — and those that were correct went to the end of the list to be asked again, while incorrect items had to be restudied (that is, the correct item was displayed on the screen for four seconds, before going to the end of the list for re-testing). Each item was pre-assigned a criterion level from one to five — that is, some words only had to be correctly recalled once or twice before disappearing, while others had to be retrieved three, four or five times.

This basic scenario was played out in two experiments. In the first, students took a recall test and a recognition test two days after the study session. In the second experiment, students were only given a recognition test (in order to eliminate any reminder effect of the recall test), and they were given it one week after the study session.

Both experiments found that higher criterion levels led to better memory.

But the effects were not precisely the same for all tests, emphasizing once again that you really do need to think about how you'll be retrieving the information 'in real life' — different scenarios require different practice, and perhaps different criterion levels.

The graph below shows the average proportion correct on the associative recognition test (in which students simply had to recognize whether the English-Lithuanian word pair was correct or not), the cued English recall test (in which students were presented with the Lithuanian word and had to type in the English word), and the cued Lithuanian recall test (students had to type in the Lithuanian word in response to the English word):

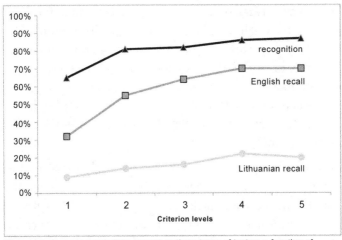

Figure 4.1: Average percentage correct on three types of test as a function of the number of times the item was correctly recalled during practice. From Vaughn & Rawson 2011.

Unsurprisingly, the best performance was on the associative recognition test. You can also see how much better performance is for the cued recall of the English words compared to cued recall of Lithuanian words. Again, this is no surprise.

But what we're interested in is which criterion level produces the best performance. See how, for cued recall of English words, having a criterion level of two (two correct retrievals) is so much better than only one, and that there's a more gradual, and even, progression for 2, 3, and 4 correct retrievals, at which point it plateaus. In other words, it appears that for this sort of situation (remembering the English meaning when presented with the foreign word), the optimum number of correct retrieval attempts on the first study session is four, but the really vital thing to do is make sure you don't stop at only one correct retrieval.

The benefit of that second retrieval attempt is also clear in the easier task of recognizing the correct English-Lithuanian pairs, and here any benefit of additional retrieval attempts is much less evident.

As with cued recall of English, the optimal number for cued recall of Lithuanian words also seems to be four, although the overall level of performance is much lower than it is for English. Given that, I can't help but wonder if the plateau at 4-5 is more apparent than real. It may be that a criterion level of 6 would produce better recall again. It also seems likely that for this much harder task, some additional help (such as the keyword mnemonic) would really help.

However, the optimal criterion level isn't only a matter of what theoretically produces the best learning. You also need to take account of the boredom factor. The returns of additional criterion levels may not be great enough to justify the increase in boredom (another reason to use something like the keyword mnemonic, which may add interest, when more practice is needed). This is important because the more boring an activity, the less likely it is that you will repeat it (or pay proper attention to it, if you do). "So much and no more" should be your watchword.

Individual items may demand different criterion levels

This principle is something you should take to heart, and apply to individual items. For example, say you were learning the following German words:

1. der Apfel

2. der Spinat

3. der Kürbis

4. der Kopfsalat

5. der Fisch

6. die Banane

7. die Karotte

8. die Kartoffel

9. die Zitrone

10. die Tomate

Some of these have very obvious English counterparts: Apfel —apple; Spinat—spinach; Fisch—fish; Banane—banana; Karotte —carrot; Tomate—tomato.

Others are less obvious: Kürbis; Kopfsalat; Kartoffel; Zitrone. But Zitrone, meaning lemon, becomes clearer once you replace the 'z' with 'c' — citron, citrus fruit, citric acid. Similarly, if you know the word 'dummkopf!', then you won't find it hard to remember that 'Kopf' is head, and the meaning of 'lettuce' for 'Kopfsalat' (salad head) is quite easy. Kürbis (pumpkin) and Kartoffel (potato), on the other hand, are much less obviously related to their English counterparts.

It seems clear, then, that some of these words are almost a case of recognition, and a criterion level of two correct retrievals should be quite sufficient when that is the case. However, more difficult items should be given a higher criterion level, such as four.

When material is very difficult, an even higher criterion level may be necessary — in which case, you should consider drafting in an additional strategy, such as a mnemonic or mapping technique, to assist.

Individual differences matter

On the subject of individual differences in learning abilities, it's worth noting that results from nearly a quarter of the students in the Lithuanian-words experiment were excluded (32 out of 131) because the students failed to reach a sufficiently high level of performance. In other words, individual differences matter! So these criterion levels are only guides — you need to work out your own optimal levels (taking into account boredom level as well as performance).

How many times should you review?

One way to cope with high criterion levels is through distribution over learning sessions, which brings us to our second question: how many times should you test yourself on the material after that first study session?

That is quite a complicated question as it interacts with another critical factor, namely spacing (which is discussed in the next chapter). But one study has given us an excellent starting point.

This study included three experiments in which students learned short texts via retrieval practice. Criterion levels varied from one to four correct retrievals in the initial session. Items also varied in how many subsequent sessions they were practiced. In these one to five testing/relearning sessions, the items were practiced until they were correctly recalled once. Memory was tested one and four months later.

The short texts contained eight key terms and their definitions (e.g., "Sensory memory is a memory system that retains large amounts of sensory input for very brief periods of time"; "Declarative memory is memory for specific facts and events that can be stated verbally"). Students were given 4 minutes to study the whole text (only some 400 words), before proceeding to the retrieval attempts. For these, students were presented with a term and had to type in their definition. Each definition contained several idea units. The students were given feedback to enable them to mark their answers.

The researchers used not only accuracy on the test as a measure of learning, but also the rate of *re*-learning — how many presentations it took for the student to re-learn an item on later review sessions. I applaud this. As the researchers remark, no one — student or teacher — expects a student to remember everything they've ever learned, particularly when (as is sadly only too common), the information is not referred to again for many months, if not years. The ease and speed with which you can refresh your memory is therefore crucial.

The aim of the first experiment was to explore the effects of initial criterion level, the question I have just discussed.

However, I'd like to briefly describe the results, partly to provide additional confirmation, but mostly because this experiment shows how initial criterion level interacts with number of study sessions. In this experiment, students were assigned a specific criterion level in the initial study session, and then participated in two review sessions: the first two days after the initial session, and the other some six weeks after that. These sessions began with a trial that served as a cued recall test.

As anticipated, at the 2-day test, recall was better the more times the item had been correctly recalled during the initial study session. See, in the graph below, how a criterion level of two produced distinctly better recall on the 2-day test than a criterion level of one, and a criterion level of three was better still.

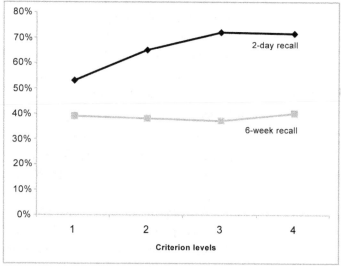

Figure 4.2: Average percentage correct 2 days and 6 weeks after initial study, as a function of the number of times the item was correctly recalled during initial study. From Rawson & Dunlosky 2011.

This confirms the 3 'norm' indicated by the previous study. Notice also that, in contrast to the previous study, these findings are very clear that going to 4 is wasted effort. I suggest that this has to do with difficulty level. The huge advantage coherent text has over foreign vocabulary is that it makes more sense. As I've discussed many times elsewhere, meaningful information is much easier to remember than relatively arbitrary information (such as individual words).

But here's what's really interesting: on the six-week test, there was no difference in recall performance as a function of criterion level (recall was around 40% at all levels). This suggests that, as long as you remember to carry out revision sessions, you don't need to worry about the number of correct retrieval attempts on the first study session.

Having said that, however, the re-learning measure did find that items that had been correctly recalled more than once were re-learned faster than words that had only been correctly recalled once, but there was no significant advantage to having a criterion level greater than two.

How did the review sessions affect learning?

The second experiment, which compared a criterion level of either one or three, and increased the number of review sessions for half the items, showed very clearly the advantage to having three revision sessions rather than just one. Average recall on the final test was less than 40% for items that only had one re-learning session, compared to over 50% for those that had three re-learning sessions. Notice, too, how poor recall was for those 'control' items that were not revised at all!

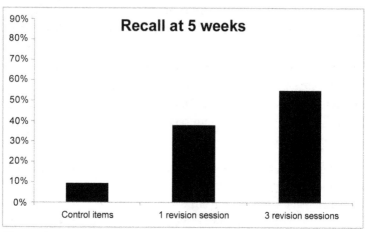

Figure 4.3: Average percentage correct 5 weeks after initial study, as a function of the number of review sessions. From Rawson & Dunlosky 2011.

Interestingly, there was also a clear benefit to having a criterion level of three compared with only one. While this benefit was greatest on the first review session (Day 3), it persisted through to the third review session (Day 10). The benefits for re-learning also continued through all learning sessions.

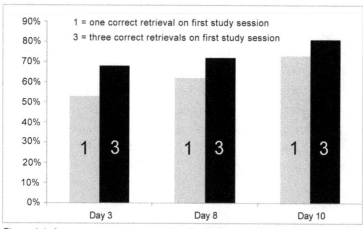

Figure 4.4: Average percentage correct on the first test trial on Sessions 2, 3 and 4, as a function of criterion level on the initial session. From Rawson & Dunlosky 2011.

The final experiment, which involved a greater number of students and more study sessions, confirmed this pattern: the benefit of having three correct retrievals rather than one was much greater on the first test, and got smaller with each subsequent session.

Session spacing may be a factor

This does, of course, make perfect sense. However, such diminishing returns may well be partly a function of the spacing. If the review sessions had been more widely spaced, the benefit — of both higher initial criterion levels, and greater number of sessions — may well have been greater.

This is something we'll look at in the next chapter.

Recommended schedule

What all this shows clearly is that having review sessions is much more important than your criterion level in that initial study session. Nevertheless, the researchers of this study recommended that **students practice recalling concepts to an initial criterion of three correct recalls and then relearn them three times at widely spaced intervals**.

One reason for this recommendation is the protection it gives if you fail to revise on schedule. Additionally, an analysis of total practice trials over all sessions (remember that on each session, the student has to correctly recall an item before it disappears) revealed that a 3 × 3 schedule required fewer retrieval attempts than other schedules producing similar levels of performance. In other words, it was the most efficient schedule if you're weighing time spent against results.

Again, it's worth noting that, across these experiments, some students' results were excluded from analysis either because they were so much better than the others or because they were so much poorer. This emphasizes that group studies can only reveal so much — they are guides, not rules. In this context, it's also important to note that these were psychology students studying psychology texts, so the experience was confounded with learning outside the experiment. Indeed, the students reported having met around half the items in class.

One final point: performance on multi-choice tests was little affected by any of this. Multi-choice tests only require recognition, not recall, which is a much easier task.

Points to remember

In your first study session, you should practice retrieving the information a certain number of times. Once is not enough! Three is a good rule of thumb.

Similarly, three is a good rule of thumb for the number of times you subsequently review the information.

Unsuccessful retrieval attempts do not count!

Because you want to avoid getting too bored, it's helpful to have some flexibility about the number of retrievals. Easy material can be limited to two successful retrievals, while harder material should have more.

As always, individual needs will vary — these numbers are a guide, not a rule.

Spacing your practice

5

> Timing is one of the key tenets of effective retrieval practice. In this chapter, I look at the amount of time you should have between initial study and review, and between subsequent reviews.

Let's begin by establishing that spaced practice is far more effective than massed practice. I'd like to think that everyone knows that, but it's an idea that has taken a while to gain traction in education, and indeed has still not changed many traditional practices that should be changed (mathematics, I'm looking at you in particular!).

The advantage of spreading out your practice

Since research has shown that most students do not appreciate the power of spacing, let's start with the 1978 study that first brought this to everyone's attention (35 years, you'd think we'd be more on board with it by now, wouldn't you?).

The study aimed to find the best way of teaching postmen to type (this was at the request of the British Post Office). The researchers put postmen on one of four schedules:

- an intensive schedule of two 2-hour daily sessions
- two intermediate schedules involving two hours a day, either as one 2-hour session, or two 1-hour sessions
- a more gradual schedule of one hour a day.

The researchers found quite dramatic differences in learning, with the 1-hour-a-day group learning as much in 55 hours as the 4-hour-a-day group in 80. Moreover, the 1-hour-a-day group remembered their skills better when tested several months later.

Here's a graph showing how well the postmen were doing after 58 hours of training, which was the point at which the 1-hour-a-day group stopped (performance is measured in terms of speed and accuracy — the measure is number of correct keystrokes per minute):

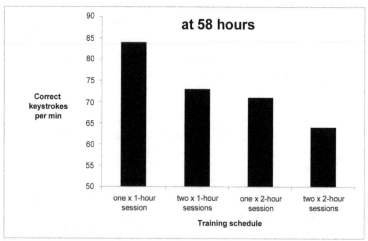

Figure 5.1: Typing speed and accuracy after 58 hours of training, comparing differently distributed sessions. From Baddeley & Longman 1978.

But (and this goes a long way to explain just why there's been so much resistance to this simple, now well-established principle), the gradual group were the least happy with the program — for although they learned much more quickly in terms of hours, it took them many more days (80 hours at four hours a day is 20 days, but 55 hours at one hour a day is 55 days).

There's an important message in that, but it's one that's more important for some people than others — for those who need quick returns to stay focused, or if the specific task is one in which you need to see quick returns. This is another aspect where you need to take into account your own personality and circumstance. However, you can take advantage of spacing's benefits without dragging things out so much.

As I've said, the evidence for the benefits of spaced over massed practice is overwhelming, and I'm not going to review three decades' worth of it. Particularly because most of it hasn't done much to tell us what we really want to know — namely, what's the best spacing for learning.

Optimal spacing

On that question, a very large study has finally given us something to work with. This study (which took place on the internet, enabling the researchers to have a very diverse range of more than 1350 participants) compared a wide range of intervals between the initial learning session and the second review session (3 minutes; one day; 2 days; 4 days; 7 days; 11 days; 14 days; 21 days; 35 days; 70 days; 105 days), and a range of intervals between the review and the test (7 days; 35 days; 70 days; 350 days).

The initial learning session involved the participants learning 32 obscure facts to a criterion level of one perfect recall for each fact. The review session involved the participants being tested twice on each fact before being shown the correct answer. Testing included both a recall test and a recognition (multi-choice) test.

What was found? Well, first of all, the benefits of having a longer space before review were quite significant, much larger than had been seen in earlier research when shorter intervals had been used. For example, if you were being tested two or three months after your review, reviewing the material three weeks after learning it would more than *double* the amount you remember, compared to reviewing it immediately after learning (this is an average, of course, and individual performance will vary).

Secondly, at any given test delay, longer intervals between initial study session and review session first improved test performance, then gradually reduced it. In other words, there was an optimal interval between study and review.

In the tables below, you can see which review interval was best for each test delay period, and how much it improved performance (compared to a review interval of three minutes). The first table shows the situation for recall; the second for recognition.

Days until test	Optimal gap before review (days)	Improvement in recall
7	1	10%
35	11	59%
70	21	111%
350	21	77%

Days until test	Optimal gap before review (days)	Improvement in recognition
7	1	1%
35	7	10%
70	7	31%
350	21	60%

See how the optimal review gap increases as test delay increases, but plateaus at certain points (this simplifies the situation of course — if you're serious about study, you're going to review it more than once!).

Note, too, that the benefits are different for recognition than for recall. For a start, and unsurprisingly, the benefits are much greater for recall than for recognition (recognition being so much easier than recall, there's much less room for improvement, especially when the test is only a week after the review).

For recognition, the greatest impact was experienced when the test was 350 days after review: reviewing at the optimal gap of 21 days produced a 60% improvement in performance — meaning that performance on the multi-choice test rose from an average of around 44% to a very respectable 70%. But the greatest benefit was seen for recall at the 70-day test delay — a dramatic 111% for those who reviewed after 21 days (an increase from a test score of around 30% to over 60%!).

Overall, and given a fixed amount of study time, the optimal gap improved recall by an average of 64% and recognition by 26%. What this means in practical terms is that if you reviewed only once, at the end of your initial study session, and ended up getting 40% on your end-of-semester exam, then, by instead reviewing the material one week after your study session, you'd boost your exam score to a very creditable 65%! A significant difference indeed. (This figure is of course an average, and I must note again that individual experiences will vary.)

Having established which of the review intervals used in the experiment were best, the researchers then estimated what the ideal intervals would really be. They came up with these:

Days until test	Optimal review gap for recall (days)	Optimal review gap for recognition (days)
7	3	1.6
35	8	7
70	12	10
350	27	25

This suggests that, if your exam is in about three months, you should review the material about two weeks after your initial study; if you want to remember the material next year, you should review it after about a month.

One caveat: while the actual results fit very nicely on the imaginary lines used to produce the estimates for test delays of 7, 35, and 350 days, the results fit less well for the test delay of 70 days. This is due to one data point: performance on the 70-day test when the review had been at 14 days was unexpectedly low. As a consequence, the discrepancy between the optimal gap produced by the participants and the estimated optimal gap is much greater for the 70-day test than it is for the other test delays — 21 days vs 12. I will return to this point below.

The need for review

Although it wasn't the focus of the study, let's take a moment to note how much recall falls after a year. In the situation where there's a 20-day gap between learning and review, those tested seven days later got pretty nearly everything right; those tested after 35 days got a very respectable average of around 80%; those tested after 70 days (about a school term) got about 60%; but those tested after a year were only hitting around 20%.

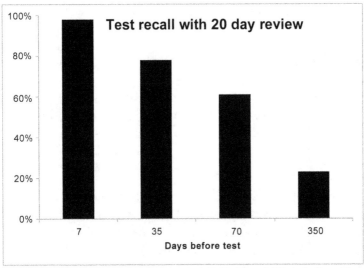

Figure 5.2: Average percentage correct on tests 7, 35, 70, and 350 days after review, when review is 20 days after initial study. From Cepeda et al. 2008.

In other words, if you don't review, or only review once, you're not likely to remember the material by next year.

This may not matter to you, of course, if you're only concerned with passing an exam and have no intention of studying any further in the area. Equally, if the information is truly important, then you would expect it to keep being used and built on, being part of a 'natural review' process. Unfortunately, this is not always (or even mostly) the case. It depends on the structure of your curriculum and how extensive your reading is. In the absence of 'natural review', more direct measures are clearly called for.

What all this emphasizes is that the common educational practice of concentrating topics tightly into short periods of time is not a strategy that is likely to produce long-lasting learning. While it might look as if you've mastered the material when you take a section-end test, chances are that you won't hold on to a lot of this information (unless you review it).

Stretching the review interval

The research also makes clear that the cost of using a review interval that is longer than the optimal gap is decidedly less than the cost of using a shorter gap — in other words, it's better to space your learning out over a too-long interval than a too-short one.

This is supported in another, more typical, study. The first experiment in this study involved students learning 40 Swahili words, and varied the gap between the first study session and the review session from 5 minutes to two weeks (and included 1, 2, 4, and 7 days). The test was ten days after the review session.

The difference between a 5-minute review gap and a one-day gap was dramatic, with recall improving from around 55% to 74%. There was little difference in having review gaps greater than a day (bear in mind the final test took place only 10 days later).

This confirms the findings of the internet study — a one-day review gap is optimal when the time until test is around a week — but this study shows a much greater benefit to that review. The difference may have to do with the learning material, which is less meaningful (and thus more easily forgotten) than the facts that were used in the large internet study.

In the second experiment, the researchers used the same sort of obscure facts used in the internet study (e.g., "Who invented snow golf?") as well as names of unfamiliar objects. The gap between the two sessions was also extended, with gaps ranging from 20 minutes to six months (including 1, 7, 28, and 84 days), and the final test given six months after the review session.

With this longer period, the best performance was achieved with a review gap of 28 days, and the decline in performance from gaps longer than 28 days was relatively small.

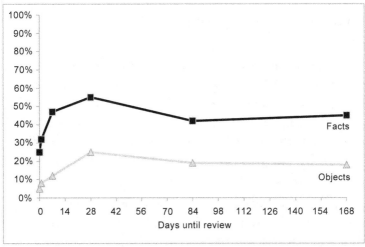

Figure 5.3: Average percentage correct for simple facts and objects, as a function of the number of days between initial study and review. From Cepeda et al. 2009.

Compare this result with those of the internet study, which

estimated an optimal review gap of 27 days when the test was 350 days later, and 12 or 21 days (extrapolated vs actual) when the test was 70 days later. This finding suggests that the longer interval is the more accurate one. But you might like to take this uncertainty as a reminder that you shouldn't rigidly regard these numbers as absolute rules! The general guide is that, if your exam is several months away, then you should review the material 3-4 weeks after studying it.

What all these experiments make clear is two general principles:

- the optimal review interval is longer when testing is delayed longer
- the penalty of stretching the interval too far is much smaller than the penalty for not stretching it far enough.

Distributing your reviews

But it's not quite as simple as saying that you should spread out your revision or practice. While the studies I've been describing have only used a single review (in order to establish a baseline), we have already established that best learning practice is to review more than once. How, then, do you space subsequent reviews? Do you spread them evenly, or at increasing intervals, or at decreasing intervals?

Expanding intervals is now considered best practice, and this idea — of progressively stretching the length of time between review so that each interval is at the limits of your memory — fits in with another idea I will discuss later, the idea of desirable difficulties. To briefly anticipate, this idea is that learning is maximized when tests are as difficult as they can be while still achieving a high measure of accuracy (i.e., keeping your errors to a minimum).

Of course, your spacing depends entirely on how well you do on your revision session! If you do well, you'll want the next interval to be longer; if you don't do well, you'll need to shorten the interval before the next session.

In other words, the key to good spacing is monitoring your learning and responding appropriately.

How type of material & task may affect spacing's benefits

Although the evidence for the value of spaced practice is considerable, there are circumstances when it doesn't seem to be so effective. The benefits are smaller with:

- meaningless material (but while meaningless material is common in laboratory experiments, it is less common in the real world)
- simple material
- incidental learning (compared to intentional learning).

In other words, spacing your practice is most important in precisely those circumstances that generally apply in study! That is, the material is meaningful and complex, and you are intentionally trying to learn it.

It has also been suggested that the spacing effect doesn't apply to inductive learning (learning to generalize from a number of examples). Reasonably enough, spaced out examples would seem to make it more difficult to notice the features the examples share. This idea brings to the fore another important factor in spacing your practice: what you do in the spaces.

We will look at this question in the next chapter.

Points to remember

Spreading your learning out (spacing) is much more effective than 'binge' learning.

The best time to review your learning depends on how long you want to remember it for.

For long-term learning, a review at around four weeks is optimal. For a semester-end exam, a week or two might be best.

Without any review (natural or deliberate), you are likely to forget almost all of the material (80%) within a year.

Best practice is to have three reviews, and these should be at increasingly longer intervals.

As a general guideline, aim for a first review one day after your initial study session, with a second review 7-10 days later, and a third review 4 weeks after that.

The best spacing depends on both you and the subject matter. As a general principle, you should aim to review at the limits of your memory — you need to find that 'sweet spot' *just before* you would forget the information. Only trial and error will teach you that!

It's better to space your learning longer than too short — make your brain work for it.

Spacing your review is of most benefit when the material you are learning is meaningful and complex.

Spacing within your study session

Timing is not only a matter of the intervals between study and review, or between reviews. It also comes into play in the distribution of items during a study session. This chapter looks at the spacing between repetitions, and in particular at what goes on in the spaces.

In the previous chapter, I talked about spacing as if the only issue was the intervals between your initial study and your reviews. However, spacing applies also to what goes on within a study session. That can be the source of some confusion. What, exactly, is meant by 'massed' or 'spaced' practice in this context?

A common procedure in experimental studies is for some words in a list to be repeated one or more times in succession, while others are repeated only after several other words have been presented. So, for example, if you're studying new vocabulary from a foreign language, you could

- test yourself on a word two or three times before moving on to the next, finishing your session when you come to the end of your words, or

- you could go through each word in the list, one by one, and then run through the list again, as many times as you think you need.

In the second case, each word is separated (spaced) from its own re-occurrence by all the other items in the list. The length of the list is therefore critical in determining spacing (this is a situation I'll discuss in more detail when I talk about practical strategies).

A third strategy, one which may reflect more common practice, is a schedule which is neither completely massed nor completely spaced. So, for example, one study presented Year 2 school-children and college students with words to be learned according to one of three different schedules:

- massed: the word was presented four times in succession

- clustered: the word was presented twice in succession and twice more in succession after eight other items had been presented

- spaced: the word was presented four times with four items between each presentation.

Filler items were presented once only.

For both the young children and the adults, learning was the same for words presented massed or clustered. Only spaced words showed better learning (see graph on opposite page).

In a follow-up experiment, Year 1 school-children were given phonics instruction in very short lessons that were either spaced (three 2-minute sessions per day) or clustered (one 6-minute session per day). After two weeks of teaching, the children were tested. Those who received the spaced lessons showed much more learning than those who had received clustered lessons: 8.3 vs 1.3 points improvement (score at final test minus score at initial test) — a significant difference indeed!

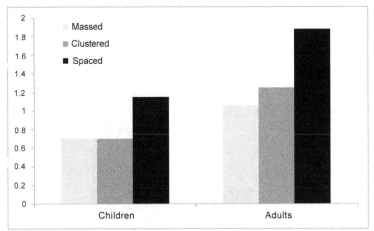

Figure 6.1: Average correct words, for children and adults,comparing presentation schedules. From Seabrook et al. 2004.

These experiments suggest the value in truly spacing individual items.

But this does beg the question, how much spacing is optimal?

I said previously, in the context of between-session practice, that it was better to have too long a space rather than too short a one. That probably applies to within-session spacing as well. But remember that optimal spacing is at that point *just before* you'd forget. Too long may be better than too short, but 'just right' is best of all.

An interesting study comparing highly supported self-study of new words with computer study may demonstrate this. The computer study used item-spacing, retrieval practice, and a criterion-level of 2 correct answers. The students were 6th and 7th graders in a poorly performing New York public school. There were four daily training sessions, with new words added each session, and old words revised. See how the students did in the first experiment, when the words to be learned were word-definition pairs (e.g., Ancestor—A person from whom one is descended; an organism from which later organisms evolved.):

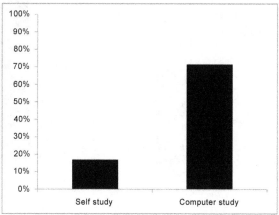

Figure 6.2: Average percentage correct on word-definition pairs, comparing supported self-study and computer-supported study. From Metcalfe et al. 2007.

Mammoth support for the benefits of appropriate practice!

But now let's look at the results of the second experiment, when the words to be learned were English equivalents of Spanish words (the children were native Spanish-speakers):

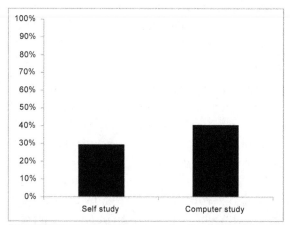

Figure 6.3: Average percentage correct on Spanish-English pairs, comparing supported self-study and computer-supported study. From Metcalfe et al. 2007.

Computerized practice is still of significant benefit, but the benefit is far smaller. Why? Simply because the words to be learned are different? The difference between word-definition pairs and Spanish-English word pairs doesn't seem great enough to produce such a massive difference in performance. Is it because there was no review session (the first experiment included a review session the day before the testing session, while the second experiment didn't). Again, this may well be a factor, but since previous words were always reviewed at each training session, it seems unlikely that the loss of a final review would have such a large effect.

No, I suggest that the principal factor at work here is the number of words to be learned — only ten in each condition in each learning session in the first experiment, compared with twenty in each condition in the second experiment. Given that the training sessions were no longer (five minutes shorter in fact), it seems more likely that the children either couldn't get enough repetition of individual items, or that the longer space between repetitions was too long for them to reliably remember them. Most likely, both are true. This may be supported by the better self-study performance: self-study's greater control of which words to study may have allowed individuals to focus only on a subset of the words, rather than try to learn all of them.

In other words, spaced retrieval practice can only do so much when you're trying to do more than you can handle.

But the study of individual words is a simple situation compared to that of studying complex information, such as the French Revolution, or photosynthesis. What does 'massed' and 'spaced' mean in that context?

Because such material is (or should be!) all connected, you need to think in terms of information-sets. Thus, spacing refers to the interval between one information-set and its review, not between your reading of each piece of information in the set. So, for example, one study used a web-based instruction module on the search for life on other planets, with one information-set describing the role of a planet's *mass* on the likelihood that a planet could be inhabited, and another information-set describing the impact of a planet's *distance* from the sun on its habitability.

When we talk about within-session spacing, the issue becomes *how* the material is spaced — what goes on in the spaces; how the items are ordered. Within-session spacing, then, inescapably brings us to **interleaving** — following an item or information-set (or problem or motor sequence) with a different item (set / problem / sequence).

The question of interleaving in cognitive practice (as opposed to skill practice) has focused on category learning, which is common in several areas of study, such as mathematics, medicine and biology. It's thought that interleaving is of special benefit to this type of learning, because category learning requires you to notice what members of a category have in common, that distinguishes them from non-members. In other words, what's important is the comparisons you can make easily. The comparisons available to you are governed by the order of items and their juxtaposition.

Note that in the context of this issue, 'massed' practice is generally referred to, for what will be obvious reasons, as 'blocked' practice.

The importance of interleaving for category and type learning

Because interleaving inevitably means items are spaced, it's been hard to disentangle the effects of interleaving from those of spacing. One study that did so, however, found a dramatic effect.

This study involved children being taught to solve four kinds of mathematics problems. In each problem, they were given the number of sides of the prism base, and were required to find the total number of faces, corners, edges, or angles by using one of the four formulas supplied. They were tested one day after the practice session.

During practice, they were given the problems either in four blocks of each type (blocked condition), or in four blocks, each of which included all four kinds of problem presented in random order (interleaved condition). In the blocked condition, brief puzzles unrelated to this type of problem were interleaved with the math problems, in order to keep the same degree of spacing in the two conditions.

Interleaving always makes the task noticeably harder (which is why students resist using this strategy), and, as expected from other research, performance during the practice session was notably worse in the interleaved condition (79% vs. 98%). However, when tested on the following day, the interleaved group performed more than twice as well as the blocked group (77% vs. 38%).

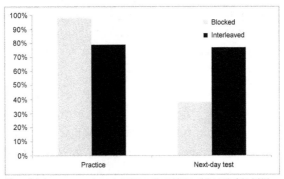

Figure 6.4: Average percentage correct on math problems, at the end of the initial practice session and on the next-day test, comparing blocked and interleaved practice. From Taylor & Rohrer 2010.

A clear demonstration of the value of interleaving (and also a very clear demonstration of why your performance at the time of learning tells you little about how well you've really learned the material). But what is it exactly that interleaving does for learning? The students were given another test, in which they were told the right formula and only had to apply it (a common situation, unfortunately, in math instruction). Here, the benefits of interleaving were not nearly as great (100% vs. 90%).

The message is clear, and it's reinforced by the type of errors made. With the exception of a single problem that was left unsolved by a single participant, every error on the test requiring the formula to be selected fell into one of two types:

- using the wrong one of the four formulas (discrimination error)

- using a completely different formula (category error).

Only one of these was affected by the type of practice. Blocked practice and interleaved practice produced about the same number of category errors (15% vs. 13%, respectively), but interleaved practice dramatically reduced the frequency of discrimination errors (46% vs. 10%).

This is, of course, not surprising: interleaving enables you to practice choosing the right formula as well as putting it through its paces; blocked practice only rehearses using the formula. (A reminder of the general principle: practice the task you need to do.)

It's clear from this why interleaving is of particular benefit to mathematics, but the broader conclusion researchers have drawn is that interleaving improves discrimination and therefore interleaved practice is probably most helpful when items or tasks are similar.

So, for example, in another study, students practiced typing three different five-key sequences on the number pad of a computer keyboard. Those who practiced the sequences in separate blocks, working a sequence until they correctly completed it 30 times (blocked practice) learned to type the sequences faster and more accurately than those in the interleaved practice condition, but this advantage of blocking was only significant for the first three blocks, and decreased steadily as time went on. Moreover, when tested the next day, those who had practiced in the interleaved condition were dramatically better than those who had practiced in the blocked condition. Average recall of the sequence was 50% for those in the interleaved group, compared to a mere 17% for the blocked group. The average timing accuracy was 83% vs 43%.

Students in the blocked group were also much worse in assessing their own learning. As we see over and over again, participants in the blocked practice condition were over-confident, expecting much better performance than they proved capable of. The interleaved group, however, were quite accurate in their predictions.

In a third study, participants were shown paintings by 12 artists and instructed to learn each one's style. In the first experiment, participants were shown all the paintings by one artist in a consecutive series (block) for six of the artists (massed condition), while the paintings by the other six artists were shown all mixed up, with participants never seeing two paintings by the same artist consecutively (spaced condition). Each painting was displayed for five seconds.

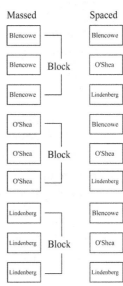

The artists whose paintings had been studied in the spaced condition were remembered markedly better than those whose work had been studied in the massed condition (an average of 61% on the test compared with 35%).

Interestingly, testing continued through four blocks, each of which showed one painting from each of the 12 artists — in other words, the test itself provided spaced learning. In response to this, the difference between spaced vs massed initial learning declined: on the fourth test block, the difference was less than 10 percentage points (around 64% vs 55%).

In a later experiment, the researchers added another two conditions:

- a variant of the spaced condition, which put cartoon drawings in the spaces (which the participants were told to ignore), so that the spacing between paintings by the same artist was the same as in the spaced condition, but all the paintings by each artist were presented in the same order as they were in the massed condition.

- a variant of the massed condition, which involved paintings by any artist appearing four at a time.

Only three artists were used. The test took place after a 20 minute filler task (watching a video).

Interestingly, these variants produced no better learning than the massed condition (all three produced average recall scores of around 60%). Improved performance only occurred when the paintings by the three artists were interleaved with each other in the standard spaced condition, with no irrelevant fillers (68%).

This supports the idea that the crucial factor for category learning is seeing the differences between the categories, and therefore what matters are strategies that make it easier to notice such differences.

To test this theory, a further experiment included a condition that presented paintings by *different* artists simultaneously. This condition maximizes your ability to notice the differences between artists, and it did indeed produce the best performance, although it wasn't significantly better than the performance produced by the standard spaced condition. However, it may suggest that, in a more demanding task, such a format might produce significantly better learning.

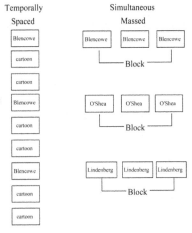

As I said, category learning is common in study. In such cases, many students are inclined to study the examples of a category all together, rather than interleaving them with other categories. Textbooks, too, tend to clump examples all together, rather than showing examples from different categories on the same page. As with massed practice, the speed with which you become fluent in processing the material gives the illusion that you have learned it quickly.

Mathematics is the prime example of this. Typically, classes and textbooks devote themselves to different kinds of problem in turn. No surprise that when the final exam comes, and students are required to choose the right method for many different problems, they often fail. A vital part of math is being able to choose the appropriate method for different kinds of problem, and yet instruction all too often fails to provide the necessary practice in this.

One study that looked at the effect of interleaving on math learning found that interleaving *trebled* test scores.

The typing sequences, the math problems, and the artists' styles, are all examples of very similar items. It may well be that interleaving is of less benefit when categories are easy to tell apart. Some studies using very simple stimuli have found massing a better strategy — but it's unlikely that you will be studying such simple information! It is the nature of complex information and skills that they share some attributes with other information-sets and skill sequences, and learning is hampered when such similarities aren't noted.

Why should interleaved practice be more effective than massed practice?

Apart from helping you differentiate skills or information-sets in terms of their similarities and differences (which is very helpful for creating good memory codes), there is another reason interleaving might be of benefit. This goes back to the general principles discussed in chapter 1.

If you keep practicing the same thing over and over again, the set of neurons encoding this new information doesn't have to be repeatedly retrieved: it just stays at a higher level of activation, hovering in the outer circle of working memory. But it's repeated retrieval from long-term memory that really strengthens a path to a memory code. Interleaving brings in more information, pushing earlier information out of working memory, forcing you to retrieve it from long-term memory.

Remember the priming effect? A code is more easily found if linked codes have recently been retrieved. If you are studying just one information-set (a list of semantically-linked words; a specific topic), then, even if you have pushed an item out of working memory, it is likely to be linked to items that are in working memory. This means the item isn't completely gone. By pushing items into working memory that are completely unrelated, you clear working memory properly of those earlier items.

Interleaving also produces more changes in context.

I said in Chapter 1 that a memory code is more easily found if the encoding and retrieval contexts match (context effect), and that a code is more easily found the more closely the retrieval cue matches the code (matching effect). This would seem to suggest that having the same context is better for learning. And yet, spacing your items out produces better learning, despite the greater changes in context produced when items are spaced.

Here we have the same sort of paradox as before: putting confusingly similar items close together produces poorer learning in the beginning, but better learning in the long run.

How do we reconcile these findings?

Learning, as I mentioned in the first chapter, isn't only about building strong paths to your memory codes. It's also about providing more paths to the codes. Repetition in the same context makes the path stronger, but multiple contexts provide more paths — and this is of increasing importance the longer the time between encoding and recall.

Let's put it this way. You're at the edge of a jungle. From where you stand, you can see several paths into the dense undergrowth. Some of the paths are well beaten down; others are not. Some paths are closer to you; others are not. So which path do you choose? The most heavily trodden? Or the closest?

If the closest *is* the most heavily trodden, then the choice is easy. But if it's not, you have to weigh up the quality of the paths against their distance from you. You may or may not choose correctly.

I hope the analogy is clear. The strength of the memory trace is the width and smoothness of the path. The distance from you reflects the degree to which the retrieval context (where you are now) matches the encoding context (where you were when you first input the information). If they match exactly, the path will be right there at your feet, and you won't even bother looking around at the other options. But the more time has passed since you encoded the information, the less chance there is that the contexts will match. However, if you have many different paths that lead to the same information, your chance of being close to one of them obviously increases.

In other words, yes, the closer the match between the encoding and retrieval contexts, the easier it will be to remember the information. And the more different contexts you have encoded with the information, the more likely it is that one of those contexts will match your current retrieval context.

Why people persist in believing massed practice is better

Unfortunately, even in the face of experience, people are remarkably resistant to changing their preference from massed to interleaved practice. Most of the blame for this lies in the fact that concentrated blocks seem to lead to much better learning at the time. The problem, that most of that learning will quickly fade, isn't obvious until later. Even when understood, this fading may well be dismissed as normal — if you don't have a comparison with more effective ways of learning, you've no reason to believe that any better learning is possible.

Sadly, the fact that blocked practice leads to better short-term performance but poorer long-term learning doesn't just fool the learners themselves, but also their teachers and instructors.

Researchers have devised a pithy term for the paradoxical effects of interleaving and spacing — **desirable difficulties**.

It seems that a certain level of difficulty during encoding leads to better learning.

On the other hand, too much difficulty can slow learning.

One of the reasons why effective studying is as much 'art' as 'science' is that it's hard to get the right level of difficulty — too simple and too hard are both, in their separate ways, potentially harmful to learning. Aptly, the benefits of getting the optimal difficulty level has been called the "Goldilocks Effect".

Interleaving is the trickiest of the strategies I mention in this book, because it does require good understanding and self-monitoring to get right. One reason is that interleaving can increase your cognitive load, giving you too much information to deal with. The trick, then, is to get the timing right. Interleaving is probably most useful when you've already achieved a certain level of competency with a skill or problem type or concept.

The other, bigger, reason why interleaving makes learning harder at the time is that it increases interference. We're all familiar with this — the way one task or piece of information can cause us to forget, or poorly remember, an earlier task or bit of information. Surely the interference between tasks or information-sets is beyond any 'desirable' difficulty?

Again, it's all about the detail.

Preventing interference

Research indicates that two skills / topics interfere with each other because the two information sets interact. This isn't simply a product of the two information sets sharing features (although the greater the similarity, the greater the interference); it's also a problem of timing.

The problem is that we're consolidating the first set of memories while encoding the second. While we *can* do both at the same time, as with any multitasking, one task is going to be done better than the other. Unsurprisingly, the brain seems to give encoding priority over consolidation (just as we so often give priority to incoming emails over the work we're in the middle of!).

Remember that new memories take several hours to stabilize. So, if you learned something a few hours ago and now you are learning something else, you are still consolidating that older learning while creating these new memory codes. What should be more reasonable than that your brain should look for commonalities between these two actions that are, as far as it's concerned, occurring at the same time? This is, after all, what we're programmed to do: we link things that occur together in space and time. Something's just happened, and now something else is happening, and chances are they're connected. So some mechanism in our brain works on that.

This can indeed be all to the good, if the two events/sets of information are connected. But if they're not, we get interference, and loss of data.

It's not only about whether or not two information-sets are consistent or not. If you think of the key problem being interference between Set 1 consolidation and Set 2 encoding, you'll see that the problem will be worse when these two processes are occurring in the same brain regions. But they don't have to. How much overlay there is will depend on what kind of material is being processed. If, for example, you learn a word list and then practice a motor sequence, the consolidation of the words will be occurring in the hippocampus while the encoding of the motor sequence will be occurring in the cerebellum. Two distinctly different (and physically very separate) brain regions.

However, if you learn the motor task first, followed by the word-list, then some interference might occur. For the motor task is being consolidated in the motor cortex, and this same region is also (somewhat surprisingly) involved in encoding words (words are tightly linked to their associated actions, faces, images, sounds, and so forth).

What all this means is that some tasks/information-sets are going to interfere with each other more than others, but it won't always be obvious why. Let experience help you determine which situations are okay and which are not, but the best general strategy is simply to provide a space for consolidation before embarking on new encoding.

Consolidation during rest

While consolidation occurs most notably during sleep, there's also evidence that a boost in skill learning can occur after rests that only last a few minutes (or even seconds). This phenomenon is distinguished from 'real' consolidation, because the gains in performance don't usually endure. However, while in some circumstances it may simply reflect recovery from mental or physical fatigue, in other circumstances it may have a more lasting effect.

One study that demonstrates this involved non-musicians learning a five-key sequence on a digital piano. In the study, the participants repeated the sequence as fast and accurately as they could during twelve 30-second blocks interspersed with 30-s pauses. A third of the participants had a 5-minute rest between the third and fourth block, while another third had the rest between the ninth and tenth block, and the remaining third had no rest at all. Everyone was re-tested the next day, about 12 hours after training.

These brief rests had a significant effect on learning, but the timing of the rest was critical.

While participants showed large improvements during training after either 5-minute rest, it was only those who were given a rest early in the training that continued to show improvement throughout the training. This group also showed the greatest consolidation gain (that is, their performance 'jumped' more than that of the other two groups when tested on the following day).

What this indicates is that consolidation is affected by the timing of the rest. Typically, in motor learning, you improve quickly in the beginning, but this rate of learning soon slows down. This pattern was indeed seen in this study — among the late-rest and no-rest groups. For these groups, improvement during blocks 4-9 wasn't as rapid as it had been during the first three blocks. Those who rested after the third block, however, didn't show this slow-down. The faster rate of learning allowed more repetition of the sequence at a higher skill level, and this may have helped develop a more stabilized memory (short-term consolidation), and thus greater overnight (long-term) consolidation.

Although having a late rest didn't benefit participants as much as having an early rest did, and the immediate post-rest jump in performance after the late rest wasn't maintained, the size of that jump did affect how much the participant's performance improved after sleep. In other words, even if you don't get the timing exactly right, brief rests are still beneficial.

Children's brains may work differently

Having said that, the story for children is different from that of adults. Intriguingly, there's evidence that consolidation and interference occur differently in pre-pubertal children.

In one study, young people (aged 9, 12, and 17) were trained on a finger-tapping task, then tested on the next two days. Some of the participants were further tested six weeks later. Another group of young students were given the same training, but also received an additional training session two hours later, during which the motor sequence to be learned was the reverse of that practiced in the initial session. They were then tested, 24 hours later, on the first sequence.

Now you'd expect, if you learned one sequence and then learned the reverse, that this would interfere badly with your memory for the first sequence. And so it did, for the 17-year-olds. But not for the 9- and 12-year-olds, who both showed the same performance gain at 24 hours that was seen when students only learned the first sequence.

Moreover, while better performance on the reverse sequence was associated with worse performance on the initial sequence at the 24-hour test for the 17-year-olds (as you'd expect), for the 12-year-olds, the better they were on the reverse sequence, the better they also did on the first sequence.

What does this mean? Why didn't interference occur in the pre-pubertal children?

One possibility is that, because their brains haven't fully developed, memory consolidation occurs differently — faster, less selectively — in children. But another reason for the lack of interference may have to do with the effects of experience. Interference only occurs when tasks overlap in some way:

spatial, temporal, conceptual, featural. If children are representing the movement sequences in a more specific, less abstract, way than adults, the sequences are less likely to share features (adults are learning a rule; children are learning two different ways of moving particular fingers). Accordingly, training on the reverse sequence provides additional training in the art of moving these fingers in this way, but doesn't interfere with the first sequence because, as far as the children are concerned, the two sequences are nothing alike.

Aging also affects consolidation & interference

Age also makes a difference at the other end of life. The brain tends to change in several important ways as we age. One of these ways is that learning gets harder. Research now suggests that a critical reason for this may have to do with consolidation.

In one study, young adults (average age 20) and older adults (average age 58) learned a motor sequence task requiring them to press the appropriate button when they saw a blue dot appear in one of four positions on the screen. The training included several learnable sequences interspersed with random trials, but participants weren't advised of this (this is a test of implicit rather than explicit learning — that is, learning you aren't consciously aware of).

As expected, younger adults were notably faster in their responses than the older group. Less expected was the fact that the older group actually learned the sequences faster than the younger group, even if they couldn't perform the task as quickly. However, on the second session a day later, while the younger adults showed the expected gain in performance from consolidation, the older adults returned to performing at the same level as they had early in the first session.

In other words, the older adults learned perfectly well during the first session, but they failed to consolidate the learning.

This pattern was confirmed in another study comparing younger and older adults, which found that, while the older adults showed improvement on an implicit sequence-learning task after 12 hours, this improvement had disappeared at 24 hours (which isn't to say that all benefit of the earlier training was lost).

Is this because we become slower to consolidate with age? This recalls the idea that children suffer less interference because they can consolidate memories more swiftly. Slower consolidation means older learning hangs around for longer, providing more opportunity for interference with later learning; faster consolidation means new information gets processed quickly, leaving less opportunity for interference.

It may also have to do with the greater interference consequent on the brains of older adults being more richly connected. Indeed, it seems likely that both processes are going on. Greater interference, and slower consolidation.

And there's a third potential factor: changes in time perception. Remember that the brain tries to associate events that occur closely in time — but what does 'closely' mean? To a child, two hours is a very long time — their brain isn't going to try and tie together events two hours apart, not without a great deal of prodding. For an older adult, on the other hand, two hours isn't significant.

Time perception is rooted in events. So one vital reason for two hours being such a long time for children is that there has been a great deal happening — a lot of new information — in that two-hour space. For an older adult, the two hours have probably been filled with the 'same old, same old' — no new information to process and encode.

In other words, interference may occur regardless of the length of time between events — what's important is what information has passed through the system in the time between.

All this suggests that interleaved practice may be even more important for older adults. And not only older adults. Although it slows down initial learning, interleaving (appropriately timed) may be especially helpful for all of those who have atrophy or impairment in specific regions (as often occurs with age, but also may happen as a consequence of injury, disease, or substance abuse).

Spacing & interleaving for complex material

But, as I said earlier, within-session spacing and interleaving are more complicated when it comes to more meaningful material. The study I referred to earlier, that used a web-based module on the search for life on other planets, found only a small benefit to interleaving, and then only for conceptual understanding (that is, learners' ability to answer novel questions involving both mass and distance concepts), not the recall of single facts or integration of facts relating to only one information-set. This speaks, I think, to the difficulty of applying interleaving to complex material.

I don't think it's surprising that presenting two closely related information-sets in an interleaved condition helped learners integrate

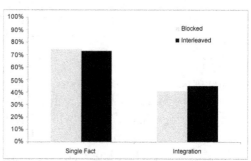

Figure 6.5: Average percentage correct on factual and conceptually integrated questions 2 days after initial study, comparing blocked and interleaved practice. From Richland et al. 2005.

them, or that whether the information-sets were interleaved or presented one after another didn't make that much difference.

Why do I say that?

What's the principal benefit of interleaving? In the examples we looked at earlier, the aim was to compare different artistic styles in order to learn how they were different; to compare different types of related math problems in order to learn when to apply different formulae; to learn different but very similar action sequences. In these types of learning, then, as I said before, interleaving assists in spotting the differences in similar tasks or types.

But here's the thing about complex information: learning it is about understanding, not memorization. Studying meaningful material, such as the causes of the American Civil War, or the conditions needed for micro-organisms to reproduce, or the meaning of Milton's *Paradise Lost*, is about learning how the information fits together. As I discuss (in great detail!) in my book *Effective Notetaking*, it's about making connections, both between all the bits of information in the material-to-be-learned, and with information you already have in your head.

I've said that, for this sort of material, interleaving and spacing should be thought of as happening between information-sets, not bits of information. But what constitutes an information-set? For of course, just as memory is a network filled with sub-networks and sub-sub-networks, right down to the very smallest memory code which is itself a network, so meaningful information is itself all connected. Where do you draw the lines, and say that *this* particular collection of information is a 'set'?

Here's yet another reason why you can't learn how to learn 'by the numbers', why you need to understand the general principles and learn how to tune in to your own learning: *you* draw the lines. And the lines will, of course, move as you learn more. This harks back to cognitive load, to the limits of your working memory and the gradual building of bigger and bigger chunks.

An information-set is something you build.

Interleaving, then, with its focus on spotting differences, would seem to be inherently unhelpful when you are building an information-set, both because of the potential interference it brings and because it adds to your cognitive load. It may be more beneficial once you have solid information-sets, during review, for example. This is something you can try if you wish, but this is an advanced strategy for those who are skilled at self-monitoring. As a general rule, it's wisest to limit the use of interleaving to skill learning (where it is very useful) and the types of category learning I've discussed.

But if interleaving is a dubious strategy for complex information, the same can't be said for within-session spacing! Remember that while interleaving inevitably includes spacing, spacing doesn't have to include interleaving. Interleaving is about mixing up tasks or information to be learned, but you don't have to keep learning in the spaces. You can fill in the spaces with inactivity, and I talked before about the benefits of brief rests to allow new information to stabilize. You can also engage in activities that don't involve new information or skills, and that also provides the brain with some mental space to process the information it's recently acquired.

There's a particularly interesting example of spaced learning in the classroom from a U.K. program developed by researchers and teachers, in which very intensive and fast-paced instruction is given in short blocks. Each instruction block is no more than 20 minutes long, and three instruction blocks are spaced by 10-minute distractor activities. There are two critical factors to this strategy:

- each of the three instruction blocks covers the same material, expressed in different ways: the first focuses on presenting the information; the second on recalling it; the third on understanding it

- the distractor activities aren't 'learning' activities, but more physical (creative or active) 'doing' activities (e.g., origami, clay modeling, ball-handling games, light aerobics).

Note the retrieval practice in the second block, while the 'understanding' block also includes recalling the information, in the context of making connections.

In one study, secondary school students (aged 13-15) were taught biology either in traditional classes over four months (a total of 23 hours instruction), or in a single Spaced Learning session. On the later GCSE test, there was no significant difference between the scores of these groups. In other words, one hour of focused, spaced, and practiced instruction, apparently produced as much learning as a term's worth of traditional classes!

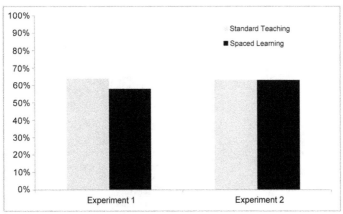

Figure 6.6: Average percentage correct on test, comparing the Spaced Learning program and traditional instruction. From Kelley & Whatson 2013.

[Note that the Spaced Learning students were disadvantaged in the first experiment, by experiencing the second biology course before the first one (traditionally taught) — this was so that any relevant prior knowledge would be minimized. In the second experiment, the courses were run in their normal order.]

In a further experiment, students experienced their normal biology classes, but were then given either the customary one-hour review of the year's material before the exam, or a Spaced Learning lesson of the same duration. Those given the Spaced Learning review scored significantly higher on the biology exam (an average of 63% vs 55%).

I'm sure you noticed how, despite the emphasis on spacing (the program is called "Spaced Learning"), the program includes several other recommendations I've discussed: variable repetition; retrieval practice; connection; time to allow stabilization of the new learning. So, this study isn't only about spacing, but is a demonstration of how these principles can be integrated and used in the classroom. Or, indeed, in private study.

Points to remember

Spacing is also beneficial within a study session.

As with between-session spacing, longer spaces are better than shorter spaces, but the best spacing captures that moment just before you'd forget.

Within-session spacing is often entangled with interleaving, and in such a case spacing depends on the number of items.

Interleaving intersperses learning of one type of task or information-set with another task or information-set.

Interleaving is of particular benefit in skill learning, math learning, and other types of category learning.

For complex material, within-session spacing can be fruitfully achieved by providing breaks.

Interleaving is probably most helpful when you need to notice differences between interleaved items.

Because interleaving makes initial learning more difficult, students usually believe that it produces poorer learning.

Interleaving and spacing help learning when they provide a level of useful difficulty. If too difficult, interleaving and spacing can harm learning.

Interleaving can increase interference, but this can be reduced by resting briefly to allow earlier learning to stabilize.

Interference is much less of a problem for pre-pubertal children, and much more of a problem for older adults.

Interleaving may be most helpful when you've achieved a certain level of mastery of individual concepts or problem types using blocked practice.

Putting it all together

Why is spaced retrieval practice so effective?

There are several ways retrieval practice may help learning:

- by matching the real testing situation, or normal-use situation
- by providing multiple routes to the memory code (by varying the contexts used in retrieval practice)
- by providing a degree of 'desirable difficulty'
- by providing opportunities to re-edit the memory code to make it easier to retrieve
- by raising your confidence in your ability to retrieve the target information.

Let's look briefly at each of these.

Matching the target situation

Remember the context principle: a code is more easily found if the encoding and retrieval contexts match.

Retrieval is always better than rehearsal, because retrieval is the task you should be practicing for, and because rehearsal gives you no feedback as to how well you've learned, while retrieval does. That's why testing is so valuable — more valuable as a learning tool than as an assessment tool. Testing *teaches*; even pre-testing (before you know the information to be learned) improves learning.

Providing multiple routes to the memory code

Every time you retrieve a memory, your starting point is a little different, and thus your path to that memory code is a little different. The more paths you have to a memory code, the greater the chance that you'll find one of those paths on any retrieval occasion. Moreover, by spacing out your retrieval practice, you increase the difference between the starting points, giving yourself a wider range of possible starting points.

By providing a degree of 'desirable difficulty'

When a task or topic seems easy, you may not give it the attention it needs for good encoding. If you read a text and it all appears very obvious, your brain probably won't even tag it as 'new'. Why, then, should it bother to encode it at all? Only by trying to retrieve it will you determine whether you do need to encode it, and doing so also tells your brain that this is something you want to remember.

Moreover, by introducing a level of artificial difficulty, through spacing or interleaving, you give your brain a reason to apply more attention to the task, with the result that you will create a stronger memory code.

Providing opportunities to edit the memory code

In the experiment comparing keyword mnemonics and

retrieval practice, keywords were remembered more often by the group that had retrieval practice, and providing the keywords on the final test significantly improved recall for the restudy-only group but didn't help the retrieval practice group. This suggests that forgetting the keyword was an important stumbling block for the restudy-only group, but the retrieval practice group forgot for different reasons. All of this indicates that the retrieval practice group remembered their keywords better. Why? A reasonable conclusion is that they had more memorable keywords — the experience of trying to recall the keywords showed them which keywords were hard to remember, motivating them to find better ones.

One of the most important benefits of retrieval practice is that it enables you to monitor your learning, and thus 'tweak' your learning strategy as needed.

Improving student confidence

When you're anxious, your worried thoughts and emotions take up valuable space in working memory, leaving less room for you to think and remember. No surprise, then, that test anxiety can significantly affect your performance on exams. By testing yourself repeatedly, you reduce that anxiety, freeing up that space in working memory for more useful thoughts.

The ten principles of effective practice

1. Practice the task you need to do.

When you are practicing or revising the skill/topic you want to learn, you should think about how you will be needing to remember this in future. For example, if you're learning foreign language words, you may wish to be able to:

- remember the English meaning of the words when faced with them.

- remember the foreign words when faced with their English counterparts.

- spontaneously generate the foreign words when talking or writing.

What you practice depends on which of these tasks you want to be able to do.

Or perhaps you're learning about the French Revolution. Your aim may simply be to pass a multi-choice test at the end of the week, or it may be to get an A on an essay-type exam in two months time, or it may be that you actually want to remember most of these details for the long-term, and be able to talk intelligently about it as you incorporate it into your broader knowledge of history.

The first step in successful learning is always to think about what you want the learning for. Only then can you work out exactly how you should be learning it.

2. The single most effective learning strategy is retrieval practice.

Most 'real' learning is aimed at being able to retrieve memories from your long-term memory store. In line with the general principle 'Practice the task you need to do', you should be practicing your retrieval of the information you want to learn.

Retrieval practice is also an absolutely critical means of monitoring your learning, which is vital if you want to learn effectively. You cannot truly know how likely you are to remember something in future if you don't test your ability to remember it.

3. When you practice retrieval, only correct retrievals count.

Indeed, because of the risk you run in learning the mistake, you may even want to add extra correct retrievals to counteract the incorrect ones. But the important thing is not to give any weight to the mistakes — do not, whatever you do, think about (or tell people) about the stupid mistake you make. If you retrieve the wrong information, mentally toss it away, turning your mind instead to the correct answer, dwelling on why that answer is correct and any aspects of it that might help you remember it.

4. Aim to do at least two correct retrievals in your first study session.

How many correct retrievals you do depends on your learning abilities, level of expertise in the topic, and personal preferences. However, having two correct retrievals, rather than only one, produces the biggest difference. Three is recommended as a general rule of thumb.

5. Space your retrieval attempts out.

There is no point in simply retrieving something that is already in working memory. To benefit from your retrieval attempt, you need to be retrieving the information from long-term memory. This means you need to have retrieved several other items (to bump the earlier item out of working memory) before trying to retrieve a particular item again.

6. Review your learning on a separate occasion at least once.

Reviewing your material at least a day later makes a big difference to your chances of remembering it in future, compared to not reviewing it, or only reviewing it at the end of your study session.

7. Space your review out.

The general principal is that the longer you want to remember the material for, the longer the length of time you should put between your initial study and your review.

However, you don't want to stretch the review interval out too far. You should aim to review the material at the point where you still remember most of it, but will soon forget it.

As a guideline:

- Review after three days if you only want to remember for a week or two.

- Review after a week if you want to remember for a month.

- Review after two weeks if you want to remember for two months.

- Review after a month if you want to remember for a year.

Note that for simplicity, this guide refers only to a single review, but for long-term learning it's best if you review more than once.

8. Review at expanding intervals for long-term learning.

If you want to remember the material 'forever', you need to review it more than once. This may be taken care of if you use the material in the natural course of your work or study. If you don't, then you need to schedule some regular reviews.

The recommended number is three, and you should aim to have increasingly longer intervals between the reviews. If you find the interval is too long and you don't remember the material as well as you'd like, then I recommend not only shortening the next review interval, but increasing the number of reviews (i.e., don't count the poor review).

If you are serious about remembering it forever, I recommend you also review the material after a year.

9. Interleave your practice with similar material.

Interleaving is particularly helpful when you are learning a skill or are doing a category learning task, such as working on math problems or learning artists' styles or plant types.

Interleaving, like spacing, introduces a desirable difficulty which forces you to pay greater attention to features that you might otherwise not notice. It is therefore particularly useful whenever there are details that you *should* notice, but probably won't.

Interleaving is also useful for ensuring that earlier items are well cleared from working memory.

10. Allow time for consolidation.

New memories take up to six hours to be stabilized, and are vulnerable to interference until that happens. You can assist that stabilization process by providing brief rests — short periods of reflection, perhaps with your eyes closed — during which you give your brain mental space to focus on consolidating the new information.

This is particularly helpful for older adults, but less necessary for children.

Specific strategies

In this chapter, I discuss different retrieval practice strategies and how to use them effectively for different subjects. The strategies include flashcards, mnemonics, Q & A, and concept mapping. 'Information learning' encompasses any sort of study topic, as opposed to skill learning (which I discuss in the next chapter).

Now that you understand why retrieval practice is so important and know the principles for practicing most effectively, let's look briefly at some different techniques for practicing.

The most obvious retrieval practice strategy — the one everyone knows — involves flashcards.

Flashcards

I'm sure all of you are familiar with flashcards, whether or not you've used them yourself. They are particularly popular for foreign language learning, but can also be used for simple facts as well as new technical words.

You can buy ready-made flashcards for some languages and subjects, and I'm not advising against this, but it is more effective if you make them yourself. Not only will the cards be customized to your own use, but the activity of selecting words and writing them down provides an additional activity that helps you learn them.

On the other hand, some of the online flashcard programs now available for language learning are very good. One of their big advantages (when done well) is in having built-in spacing and interleaving. The downside is that the spacing is rarely customizable and may not be right for your individual needs.

To keep it simple, in this discussion I only talk about the traditional format of physical cards (which have, indeed, some advantages over a virtual program). The basic principles, however, can be applied to virtual cards.

A standard way of using flashcards is simply to go through a set number each day, separating out those you have trouble with, so you can review them more often. For example, if you keep troublesome ones handy, you can go through them at odd moments during the day when you're waiting for something.

Flashcards aren't simply useful as a means of testing yourself. For example, if they're word (as opposed to fact) cards, you can use them to group words in different ways, play a variety of card games with them, or make a bingo game with them. However, since our focus is practice, I'm going to restrict myself to discussing the best way to use them in their standard way: by going through a pile of cards, looking at one side, and trying to retrieve the matching response.

There are two main questions when it comes to using flashcards (ignoring the spacing and distribution questions, which I have already covered in earlier chapters):

- When should you 'drop' a card?
- How many cards should you have in a stack?

Let's start with dropping.

When to drop a card from the stack

Now, dropping cards from your stack that you have 'learned' does make sense, because it gives you more opportunity to practice the harder cards. The big problem is knowing when it's the right time to drop them. A study that looked at this found that people dropped cards for two reasons: the obvious one, that they thought they knew them and didn't want to waste time on them, and a less obvious one — that they considered these cards too hard to learn.

There's nothing especially wrong with these principles, but in both these cases, the students' judgment tended to be awry. That is, they dropped easy cards too soon (remember how much difference that second correct retrieval makes), and they had insufficient faith in the power of retrieval practice, not realizing that if they had persevered with the harder cards, they would, for the most part, have found them learnable.

Aside from not realizing the power of retrieval practice, the students' major problem lay in not knowing this learning principle: **the value of studying is highest when items are closest to being learned**. Too often, students, fooled by the easy fluency that comes from repeating items in quick succession, believe they have learned something before they truly have, and stop practicing something just at the moment when studying is most valuable.

There's another problem with dropping — by doing so, you reduce the spacing of the cards you have left in the stack, which are presumably the more difficult cards. By having them come up more quickly, they will become more 'massed' than 'spaced', making you more likely to be fooled by the fluency effect into thinking you have learned them before you truly have.

Now, none of this is to say you shouldn't drop cards that you've learned. But you do need to be aware that it's not all benefit. The best strategy is to guard against dropping cards too soon. As a rule of thumb, I'd suggest that, when you get to the point where you think you can drop a card, give it one more turn. Most importantly, *don't* drop a card after only one correct retrieval!

Also, resist dropping too many too soon. If there are so many easy cards in your stack, consider adding more cards to the stack. Remember, to be effective, you don't want cards turning up again too quickly.

Relatedly, and more importantly, this dropping should not be carried over to the next time you run through the stack. Every time you review, all the cards should be returned to the stack, although you can drop the easy cards more quickly on these reviews.

How many cards in a stack

The question of how many cards there should be in a stack was explored in a study using GRE-type word pairs, such as *effulgent: brilliant*. Using digital flashcards, students studied two 20-word lists. For one list, the words were shown in a single stack of 20 flashcards, which were presented four times, always in the same order, and for the other list, the words were shown in four stacks of five cards, with each stack presented four times before going on to the next stack. With the single stack, then, each card only came around after 19 other cards, while with the smaller stacks, each card came around after only 4 cards.

Cards in the more spaced condition were remembered significantly more when tested 24 hours later — 49% vs 36%.

As usual, after the learning sessions, students anticipated that they would do notably better with the smaller stacks than with the large stack. This is consistent with common student practice, which appears to be to divide larger amounts of cards into smaller stacks. Indeed, a leading GRE study guide apparently advises this (or did at the time this research came out, in 2009).

The next experiment employed a more realistic study situation, reflecting common student practice, and found the benefit of a large stack over smaller stacks became dramatic. In this scenario, the students spread their study over four days, with the single large stack being gone through twice on each day, and the small stacks each being reviewed eight times on one of the days. In other words, each card, regardless of condition, was experienced eight times.

In this situation, the large stack was recalled at an average rate of 54%, vs a mere 21% for the small stacks!

A further experiment had the students review all their cards on the fifth day, before being tested on the following day — again, a more realistic scenario. Although this improved performance on the test, the marked difference between the conditions was maintained, with an average correct for the large stack of 65% vs 34% for the small stacks.

In other words, having a larger stack nearly doubled test scores.

The researchers also analyzed individual differences, finding that spacing was more effective for 63 of the 70 students (90%), with three doing equally well in both conditions, and four doing better in the massed condition. Another reminder that general guidelines are only that, and even when a strategy is as strongly supported as this, a few individuals will always find a different approach more effective. You must work out what works best for you.

> ## Best practice for flashcards
>
> Don't drop a card before having at least two successful retrievals.
>
> Use a reasonably large stack, in which cards are interspersed by a goodly number of other cards. Twenty cards is probably a good starting point — increase or decrease this in response to your performance.
>
> Don't drop a card permanently — return it to the stack for the next review.

Flashcard variant

An easy variant for those who don't want to go to the trouble of making flashcards is simply to take a lined sheet of paper and fold it lengthwise down the center. Write the words or questions down one half, and write the meanings/English translation or answers down the other.

This is much easier (and cheaper) than writing a lot of flashcards, but does of course come with significant drawbacks. In particular:

- you can't change the order so readily (you can randomly skip all over the list, but this does make it harder to ensure that you've done them all)

- you can't group them in different ways

- you can't as easily drop items as you go (although you can make check marks beside the items and use these as a guide, which does have the advantage of providing a record of exactly how well you've done on previous occasions)

- you have to wait until the end of the page before you check any answer (or, if you do check an answer before you're finished, you'll see some of the other answers).

Nevertheless, using this technique may be sufficient in certain circumstances — for example, if you're cramming for a test, or if you want to do an initial cull and see which items are more resistant to learning.

Keyword mnemonic

Like flashcards, the keyword mnemonic is especially popular as a means of memorizing new vocabulary, particularly for foreign languages. It has also proved its effectiveness for learning other associated pairs, such as authors and books, paintings and artists, countries and capitals, minerals and their attributes.

As I said earlier, this method involves choosing an intermediary word that binds what you need to remember to something you already know well. It's usually recommended that the word be something concrete that can easily form an image, although this isn't an absolute requirement. Sometimes it's very difficult to find an appropriate visualizable word and you must settle for a more abstract one. Some people, moreover, find it much easier to create verbal connections than visual ones.

Here's an example of how the keyword mnemonic can be used to remember a simple fact: that Canberra is the capital of

Australia. You could create an image involving a *can* of *beer* (an obvious phrase for Canberra, particularly in light of the Australians' notorious enjoyment of beer!), and an Australian icon such as a kangaroo or a koala bear (to signal "Australia"). Thus, your image for remembering this fact could be a kangaroo swigging back a can of beer.

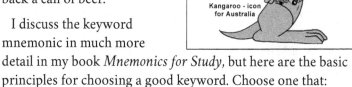

I discuss the keyword mnemonic in much more detail in my book *Mnemonics for Study*, but here are the basic principles for choosing a good keyword. Choose one that:

- sounds as much as possible like the to-be-learned word

- easily forms an interactive image with the word meaning

- is sufficiently unlike other keywords you're using to not be confused

- is a familiar word, easily recalled.

And here are the basic principles for constructing the image. An effective image:

- connects the keyword with its associated word (usually the target word's meaning) in a way that is *active* and personally *meaningful*

- is simple

- is clear and vivid.

Note that the really important principle is that the image be *interactive* — that is, a kangaroo *drinking* a can of beer, not a picture of a kangaroo *and* a beer can; a letter *in* a cart, not a letter *and* a cart.

The keyword mnemonic is an adjunct strategy in the context of practice. Flashcards are a retrieval practice strategy; mnemonics, on the other hand, are strategies for transforming hard-to-remember information into something more memorable. In other words, their role here is to provide a different content for practice. Accordingly, and given the power and ease of retrieval practice alone, I recommend that you limit your use of mnemonics to items that are not easy to learn without additional help.

The keyword mnemonic is one of the easier mnemonics to master, nevertheless, you may be put off because it requires a certain amount of creativity. Do be assured that coming up with words and images gets easier with practice.

The keyword method is undoubtedly an effective learning strategy, but it is mainly a strategy for recognition learning. You see the word *carta*, the keyword *cart* is triggered, and hopefully the image of the letter in the cart is then recalled and you realize the meaning of *carta* is *letter*. You see the word *Canberra*, you think of the kangaroo drinking the can of beer, you realize that the kangaroo indicates Australia. The method is not, unfortunately, as useful the other way around, that is, for remembering the Spanish for *letter*, or remembering the capital of Australia. But you can help with that by making sure you do your retrieval practice in both directions.

One way in which many users fall down is in focusing on the wrong aspect of the mnemonic. It's natural enough to concentrate on retrieving the word to be learned, but in fact it's better to focus (not exclusively!) on the link between the keyword and the image (or verbal counterpart), rather than the link between keyword and word-to-be-learned. The link between keyword and image is vital, because it's the image that holds the link to the target word. Don't assume you'll remember it without specific practice.

So, for example, you need to practice recalling the image of the letter in the cart in response to *carta*, and recalling the image of a letter in a cart in response to *letter*.

Indeed, I'd like to emphasize that the relationship between mnemonics and retrieval practice works both ways. It is not simply that the keyword mnemonic is a useful adjunct strategy to retrieval practice; retrieval practice is a critical adjunct to using mnemonics successfully.

This is something that often gets overlooked in mnemonics training programs. But the purpose of mnemonics is to make information more memorable, not to make it so memorable that you never need to practice it! Moreover, not only do novices tend to under-practice their mnemonics, but they often practice the wrong thing. Whatever the mnemonic (and there are several mnemonic strategies, which I cover in *Mnemonics for Study*; I have focused here on the keyword mnemonic because it has the broadest and clearest benefits), you need to practice all the links in the mnemonic chain.

Using mnemonics for complex information

To take a quite different mnemonic, let's consider the situation in which you have to give a speech without notes. In such a situation you want to chain your cues, so that each one recalls the next one in the sequence. Each cue also needs to elicit all of its associated information. So first you need to practice recalling the information associated with each cue, and then you need to practice recalling each link in the chain. This is not quite as straightforward as it may seem.

Here, for example, are the main points you might want to cover if giving a speech on the genesis of suicide terrorism (based on an article by Scott Atran, published in *Science*, and reproduced in

The Best American Science and Nature Writing 2004):

- Definition (freedom fighters; French Resistance; Nicaraguan Contras; US Congress, act; two official definitions; restriction to suicide terrorism)

- History (Zealots; hashashin; French Revolution; 20th century revolutions; kamikaze; Middle East – 1981 Beirut; Hezbollah; Hamas; PIJ; Al-Qaida - Soviet-Afghan War; fundamentalism error)

- Difficulties of defending against (many targets, many attackers, low cost, detection difficulty; prevention)

- Explaining why (insults; attribution error; Milgram; perceived contexts; interpretation)

- Poverty link (crime – property vs violent; education; loss of advantage)

- Institutions (unattached young males, normal, personal identity – Palestinians, Bosnians; peer loyalty; emotional manipulation)

- Benefits (to individuals, to leaders, to organizations; effect of retaliation)

- Prevention strategies (searches; moles; education; community pressure; need for research)

Now, if you simply learn this chain of cues, perhaps through a story/sentence mnemonic, such as "**Definitions** of **history fail** when they don't **explain why poverty is dangerous, how institutions benefit**, and how to **prevent** this", then you're going to have to mentally recite this each time you come to the end of a section and need to move on to the next one. You'll also have to recall which cue belongs to the section you've just finished before you know which one is next.

Rather than rehearsing your chain of cues, then, it's better to practice finishing each section with a sentence that recalls the cue (for example, at the end of your section on history, you say "Well, that's the history."), and practice each link as an associated pair. Thus, **history** is linked to **difficulties** — you practice retrieving "difficulties" when you hear "history".

Practicing this in variants may help you not get fixated on reproducing your text verbatim, and protect you from being knocked off course by some slight difference in your words. So you could practice: "Well, that's the history. Let's talk about the difficulties of defending against this type of terrorism."; "So much for history. Let's talk about the difficulties of defending against suicide bombers." ; "I hope this recounting of the history of suicide terrorism gives you some idea of the difficulties of defending against this type of terrorism. Let's talk about that next."; and so on.

In the related case of writing an essay in an exam situation, however, it's fine to simply practice the complete chain, because you can look back over what you've written. You can even jot down the whole chain at the beginning.

The keyword method can also be used for several items of related information, such as the main points in a text. If you do this however, you should aim to create one single integrated mnemonic rather than lots of separate ones. So, in the above example, you would want to create a single integrated mnemonic for "Definition (freedom fighters; French Resistance; Nicaraguan Contras; US Congress, act; two official definitions; restriction to suicide terrorism)" and for each of the other bullet points, plus a single integrated mnemonic for the whole thing (Definition; History; Difficulties; Why; Poverty; Institutions; Benefits; Prevention). Having said that, I wouldn't particularly recommend the keyword method for this situation, which requires you to remember order. A number of mnemonic strategies are designed to help you remember a specific order; the keyword mnemonic is not one of them.

A more appropriate situation for an integrated keyword mnemonic would be something like this, showing that the plant classification *Angiosperms* contains the class *dicotyledons*, which in turn contains the three orders *rubiales*, *sapindales*, and *rosales* (example taken from Carney & Levin, 2003):

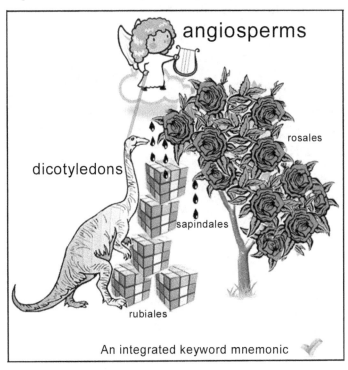

An integrated keyword mnemonic

In this integrated keyword mnemonic, the keyword 'angel' represents *angiosperms*; the keyword 'dinosaur' represents *dicotyledons*, 'roses' for *rosales*, sap for *sapindales*, and 'Rubik's cubes' for *rubiales*. The picture shows the *angel* overseeing its pet *dinosaur* (note the leash), that is climbing the pile of *Rubik's cubes* so that it can lick the *sap* dripping down from the *rose* tree.

You may think that this would be much more difficult to remember than several simple mnemonics, but the big advantage of a truly integrated mnemonic (and it must be integrated, with all the parts forming a whole, not simply a collection of images, such as the image below) is that it tells a story. As I said earlier, the human brain loves stories, and finds them much easier to remember than isolated facts.

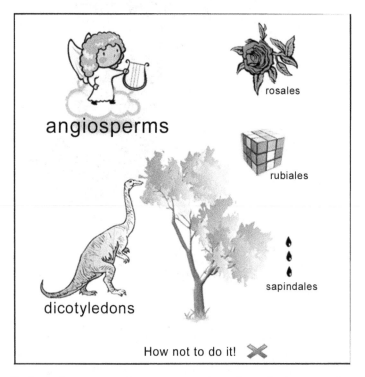

How not to do it! ✖

So how do you practice a mnemonic such as the integrated keyword mnemonic above?

Unless you are skilled at sketching, you probably don't want to practice drawing the mnemonic (if you are, go for it!). But, in general, the idea of using imagery mnemonics is not to physically draw the images (although an initial image can help a lot, particularly if you are as yet unskilled at mental visualization). The idea is to form, and practice, mental images — the more vivid the better. In the beginning, you probably want to practice recalling to mind your image in response to the most important retrieval cue. Thus, in this case, you would use the angel to anchor your image, as it represents the overarching concept (angiosperms).

Let's do a mental run-through of the process.

"What type of plants do angiosperms include?"

"Angiosperms — angel. Okay, I've got an angel ... and it's holding a leash on its pet dinosaur. Dinosaur — dicotyledon. The dinosaur's climbing up ... a stack of Rubik's cubes. Rubik — rubiales. Okay, angel (angiosperm), dinosaur (dicotyledon), Rubik's cubes (rubiales). And there's roses ... it's a big rose tree ... and the roses are dripping sap. That's why the dinosaur's climbing the Rubik's cubes, to get the sap from the roses. Roses — rosales. Sap — sapindales. So ... angiosperms include dicotyledon, and they include rubiales, rosales, and sapindales."

Notice how I've spelled everything out. That isn't simply for explanatory purposes. As I said before, novices tend not to focus on all the links in the mnemonic chain, but until you've got it well mastered, you really do want to make sure you check every link. The more tightly connected each item is, the more likely it is that you will retrieve the whole image, and the more opportunities you will have to get at it — for example, with a retrieval cue of dicotyledon, or sapindales.

The other thing to note is that while you're going through this explicit verbalization of each link in the chain, you are simultaneously trying to visualize the images as vividly as you can.

The process will of course get quicker each time you retrieve the image, and eventually, if you have given it sufficient practice, the mnemonic itself won't need to be explicitly called. As with a skill, which starts off being explicitly verbalized, and later becomes a memory that you seem to have encoded in your body, the mnemonic becomes unnecessary once it has served its purpose, and you should just 'know' that angiosperms include dicotyledons which include rubiales, rosales, and sapindales.

Using a mnemonic effectively

- only use a mnemonic when the material cannot be meaningfully linked, or must be remembered in a specific order

- limit your use of mnemonics to material that you need extra help to remember

- always think about how you'll be trying to remember the material

- focus on every link in the mnemonic chain.

Questioning

Flashcards and mnemonics are tools for learning words and simple facts. Although these are valuable details that can anchor more complex material, your study of meaningful material requires another approach. Earlier, I discussed how you need to produce a set of learnable points from such texts (or lecture notes), and talked blithely about turning these into a 'Q & A' format. While this should be a relatively straightforward task (by far the trickier part is selecting the important points in the text), here are some pointers to the process.

Let's use the set of learnable points from the ozone text to demonstrate the process in action. This is a more complete learnable set, including information omitted earlier, when I assumed some background knowledge.

1. Ozone is important because it shields the surface from harmful ultraviolet radiation.

2. The stratosphere holds 90% of the ozone in our atmosphere (the ozone layer).

3. The troposphere holds 10%.

4. The troposphere is the lowest part of our atmosphere, where all of our weather takes place.

5. The ozone layer protects us; tropospheric ozone is a pollutant found in high concentrations in smog.

6. Wavelength is a measure of how energetic is the radiation.

7. The visible part of the electromagnetic spectrum ranges from 400 nanometers to 700 nm. Red light has a wavelength of about 630 nm; violet light about 410 nm.

8. Radiation with wavelengths shorter than those of violet light (at the short end of the visible spectrum) is called ultraviolet radiation. UV waves are dangerous because they're energetic enough to break the bonds of DNA molecules.

9. Of the three different types of ultraviolet (UV) radiation, the shortest (UV-c) is entirely screened out by the ozone layer, while the longest (UV-a) is not so damaging, so the main problem is UV-b.

10. The high reactivity of ozone results in damage to the living tissue of plants and animals, and is often felt as eye and lung irritation.

11. While our bodies can repair the damage done by UV waves most of the time, sometimes damaged DNA molecules are not repaired, and can replicate, leading to skin cancer.

12. The strong absorption of UV radiation in the ozone layer reduces the intensity of solar energy at lower altitudes. More energetic photons (ones with shorter wavelengths) are also less common.

13. Because ozone is most protective on the most dangerous wavelengths, a 10% decrease in ozone would increase the amount of DNA-damaging UV by about 22%.

14. Time and season affect how much UV radiation is absorbed by ozone because the angle of the sun affects how long the radiation takes to pass through the atmosphere (the path is shorter when the sun is directly overhead, so the radiation meets fewer ozone molecules).

15. Measurement:

  a) Solar flux = the amount of solar energy in watts falling

perpendicularly on a surface one square centimeter; units are watts per cm^2 per nm.

b) The action spectrum measures the relative effectiveness of radiation in generating a certain biological response (such as sunburn) over a range of wavelengths.

You will learn more from this if you try turning these into questions before looking below at my suggestions.

Some of the learnable points are very simple and readily break down into a Q & A set. It's useful to include alternate ways of expressing the questions.

1. Ozone is important because it shields the surface from harmful ultraviolet radiation.

   Q: Why is ozone important?

   A: Because it shields the surface from harmful ultraviolet radiation.

   Q: What shields Earth's surface from harmful ultraviolet radiation?

   A: Ozone

2. The stratosphere holds 90% of the ozone in our atmosphere (the ozone layer).

   Q: What proportion of the atmosphere's ozone is in the stratosphere?

   A: 90%

   Q: Where does 90% of atmospheric ozone lie?

   A: The stratosphere.

3. The troposphere holds 10%.

Q: What proportion of the atmosphere's ozone is in the troposphere?

A: 10%

Q: Where is ozone found in the atmosphere, and in what quantities? (This merges both Q2 and Q3.)

A: 90% in the stratosphere and 10% in the troposphere.

4. The troposphere is the lowest part of our atmosphere, where all of our weather takes place.

Q: What part of the atmosphere is the troposphere?

A: The lowest part, where weather happens.

Q: Where in the atmosphere does weather occur, and what is this region called?

A: In the lowest part, the troposphere.

6. Wavelength is a measure of how energetic is the radiation.

Q: What does wavelength tell us?

A: How energetic the radiation is.

Q: What property of matter shows how energetic its radiation is?

A: Wavelengths.

Note that even simple learnable points can sometimes break down into several questions:

7. The visible part of the electromagnetic spectrum ranges from 400 nanometers to 700 nm. Red light has a wavelength of about 630 nm; violet light about 410 nm.

Q: What is the range of the visible part of the electromagnetic spectrum?

A: 400 nanometers to 700 nm

Q: What is the wavelength of red light?

A: About 630 nm.

Q: What is the wavelength of violet light?

A: About 410 nm.

Sometimes you have to pull out what is 'between the lines' (i.e., not explicit), to seek out the important message. For example, from the following learnable point you could offer a series of straightforward questions:

5. The ozone layer protects us; tropospheric ozone is a pollutant found in high concentrations in smog.

Q: What does the ozone layer do?

A: The ozone layer protects us.

Q: What does tropospheric ozone do?

A: Tropospheric ozone is a pollutant.

Q: Where is this pollutant found?

A: In smog.

These might, in fact, be what you need, if you are coming to the topic with no relevant knowledge at all. But if you know well enough that the ozone layer protects us, it's probably better to go with the following question (note that your aim should not be to make up easy questions for yourself! Your learnable points should encompass the important information that you don't know, or don't know well enough, and your questions should reflect that):

Q: Is ozone always protective?

A: No. The ozone layer protects us, but tropospheric ozone is a pollutant found in high concentrations in smog.

So, with the following learnable point, you might feel the need to include three questions:

8. Radiation with wavelengths shorter than those of violet light (at the short end of the visible spectrum) is called ultraviolet radiation. UV waves are dangerous because they're energetic enough to break the bonds of DNA molecules.

   Q: Where is violet light in the spectrum?

   A: At the short end (i.e., the end with the shorter wavelengths).

   Q: What is ultraviolet radiation?

   A: Radiation with wavelengths shorter than those of violet light.

   Q: Why is ultraviolet radiation dangerous?

   A: Because it's energetic enough to break the bonds of DNA molecules.

When you break points down to questions in this way, you might also become aware of additional questions not answered in your learnable points. So, for example, you may not know what "break the bonds of DNA molecules" means. You may need to return to the text, or perhaps even have to seek further to answer your questions. Whether or not you do so depends on your motivation and goals.

The next two sets of questions may similarly make you realize there's some confusion in your mind.

9. Of the three different types of ultraviolet (UV) radiation, the shortest (UV-c) is entirely screened out by the ozone layer, while the longest (UV-a) is not so damaging, so the main problem is UV-b.

Q: How many types of ultraviolet radiation are there and what are they?

A: Three. UV-a, UV-b, and UV-c.

Q: Which of the three different types of ultraviolet radiation is most dangerous and why?

A: UV-b. Because the shortest (UV-c) is entirely screened out by the ozone layer, and the longest (UV-a) is not so damaging.

Again, you may want to pursue the question of why the shortest wavelengths can't get past the ozone layer, and why the longest wavelengths aren't as damaging.

10. The high reactivity of ozone results in damage to the living tissue of plants and animals, and is often felt as eye and lung irritation.

Q: How does ozone damage us?

A: Because ozone is a highly reactive form of oxygen, it can irritate the eyes and lungs.

This begs the question of what "highly reactive" means. Again, you may understand what it means, or you may not care.

Some statements are less readily turned into questions. The following, for example, requires a little thought. Your first off-the-cuff idea may be something like "How do UV waves damage us?", but is this really the best question? Sometimes it helps to think of the answer first, and work back. In this case, the "While" is making the Q & A format less obvious. Try ignoring that to start with.

11. While our bodies can repair the damage done by UV waves most of the time, sometimes damaged DNA molecules are not repaired, and can replicate, leading to skin cancer.

A: Skin cancer occurs when damaged DNA is not repaired.

Having created this answer, the question is more obvious.

Q: Why does UV damage sometimes cause skin cancer?

And now, in response to that question, you may want to expand on the answer, incorporating what was alluded to in that "While ...".

A: Skin cancer occurs when DNA damaged by UV radiation is not repaired, as it usually is.

The following question is in a format that is misleadingly easy to transform into a Q & A:

12. The strong absorption of UV radiation in the ozone layer reduces the intensity of solar energy at lower altitudes. More energetic photons (ones with shorter wavelengths) are also less common.

Q: What does the strong absorption of UV radiation in the ozone layer do?

A: Reduce the intensity of solar energy at lower altitudes.

When framing questions from a text, you know what the

answer is meant to be. That's fine as long as the text is fresh in your mind, but if you're seeking longer-lasting memories you want questions that are phrased in such a way that someone knowing the topic but not the specific text would answer it 'correctly' (i.e., how you meant it to be answered). The following version makes it clearer what information you're seeking:

> Q: What effect does the ozone layer's absorption of UV radiation have on energy levels at lower altitudes?

An alternate question that gets to the heart of the learnable point without being so 'leading' would be:

> Q: Why is it safer for living things at lower altitudes?
>
> A: Because the ozone layer absorbs a lot of the UV waves, reducing the sun's intensity, and because the more energetic photons are less common.

Sometimes you need to refer back to earlier points to properly answer the question. In the next learnable point, the "most dangerous wavelengths" are not explicitly identified, but adding it to your answer does reinforce an important point (that UV-b are the most dangerous wavelengths).

13. Because ozone is most protective on the most dangerous wavelengths, a 10% decrease in ozone would increase the amount of DNA-damaging UV by about 22%.

> Q: Which wavelengths does ozone protect us from most?
>
> A: The most dangerous wavelengths: UV-b
>
> Q: How much would a 10% decrease in ozone increase the amount of DNA-damaging UV?

A: By about 22%.

A "because" in a learnable point makes the Q & A pretty obvious:

14. Time and season affect how much UV radiation is absorbed by ozone because the angle of the sun affects how long the radiation takes to pass through the atmosphere (the path is shorter when the sun is directly overhead, so the radiation meets fewer ozone molecules).

    Q: Why does time of day and season affect how much UV radiation is absorbed by ozone?

    A: Because the angle of the sun affects how long the radiation takes to pass through the atmosphere (the path is shorter when the sun is directly overhead, so the radiation meets fewer ozone molecules).

Note that this could have been broken into two questions, rather than adding that explanatory note in brackets in the answer. The issue of how many questions you use is reminiscent of the issue of how many steps you break a process into — it depends on your 'chunks', that is, on your existing knowledge and your grasp of the material.

Definitions, as in the last two points, would seem to be readily transformed into a Q & A (What is a ....? What does .... measure?), but there are two aspects you might like to note. One is whether you need an additional question relating to the unit of measurement, and the other is whether you want to pursue further explanations of any unclear concepts. For example, what does "the relative effectiveness of radiation in generating a certain biological response over a range of wavelengths" really mean?

15a. Solar flux = the amount of solar energy in watts falling perpendicularly on a surface one square centimeter; units are watts per cm2 per nm.

Q: What is solar flux?

A: The amount of solar energy in watts falling perpendicularly on a surface one square centimeter.

Q: What unit is solar flux measured in?

A: watts per cm2 per nm

or:

Q: How is solar energy measured?

A: In watts falling perpendicularly on a surface one square centimeter.

15b. The action spectrum measures the relative effectiveness of radiation in generating a certain biological response (such as sunburn) over a range of wavelengths.

Q: What does the action spectrum measure?

A: The relative effectiveness of radiation in generating a certain biological response (such as sunburn) over a range of wavelengths.

Q: What is used to measure the capacity of ultraviolet radiation to damage living tissue?

A: The action spectrum

One important aspect of using a Q & A format for retrieval practice is the issue of organization and order. I have said that practicing in different ways is best because it provides opportunities for different retrieval cues to be used, new paths to be forged. However, in the case of meaningful information, this strategy should be tempered. Remember that your priority is to create and strengthen good memory codes — ones where the various bits of information are tightly clustered and strongly connected. The human brain loves stories. Stories form just about the best links there are, and one of the reasons for this is that stories ('proper' stories!) are based on a narrative chain — a cause-&-effect chain.

You don't need narrative to have a chain of linked events. A lot of science, and history (of course), is explained in such a way. You want to take advantage of this, which means you don't want to mess with the order too much, when studying or revising the material.

This is particularly true in your early reviews. Initially, you do want to concentrate on getting that integrated story/network firmly into your head, so it's best to practice the facts in logical order. Once you've got that, you can jumble up the questions as thoroughly as you like, to build up different paths and strong retrieval cues. What you can do, however, to provide some useful variety is to have alternative questions, such as I provided for some of the above examples.

Do note that 'logical order' doesn't necessarily mean the exact order of points in the text. Move them around to whatever makes sense to you. Here, for example, is a comparison of the questions in the original order of the text and in an order that's logical for me (to keep this list from becoming too long, I haven't included alternative questions):

Note that, when you re-organize them, and when you only focus on the questions, you may realize that a particular question only made sense as a follow-up to the answer of its original predecessor (for example, Question 11). This suggests a re-phrasing of the question, perhaps even of both questions.

Original order	Practice order
1: Why is ozone important?	1: Why is ozone important?
2: What proportion of the atmosphere's ozone is in the stratosphere?	9: What does the ozone layer do?
3: What proportion of the atmosphere's ozone is in the troposphere?	2: What proportion of the atmosphere's ozone is in the stratosphere?
4: What part of the atmosphere is the troposphere?	3: What proportion of the atmosphere's ozone is in the troposphere?
5: What does wavelength tell us?	4: What part of the atmosphere is the troposphere?
6: What is the range of the visible part of the electromagnetic spectrum?	12: Is ozone always protective?
7: What is the wavelength of red light?	10: What does tropospheric ozone do?
8: What is the wavelength of violet light?	11: Where is this pollutant found?
9: What does the ozone layer do?	17: How does ozone damage us?
10: What does tropospheric ozone do?	5: What does wavelength tell us?
11: Where is this pollutant found?	6: What is the range of the visible part of the electromagnetic spectrum?

Original order	Practice order
12: Is ozone always protective?	7: What is the wavelength of red light??
13: Where is violet light in the spectrum?	8: What is the wavelength of violet light?
14: What is ultraviolet radiation?	13: Where is violet light in the spectrum?
15: Why is it dangerous?	14: What is ultraviolet radiation?
16: Which of the three different types of UV radiation is most dangerous and why?	15: Why is it dangerous?
17: How does ozone damage us?	16: Which of the three different types of UV radiation is most dangerous and why?
18: Why does UV damage sometimes cause skin cancer?	21: Which wavelengths does ozone protect us from most?
19: What does the strong absorption of UV radiation in the ozone layer do?	18: Why does UV damage sometimes cause skin cancer?
20: Why is it safer for living things at lower altitudes?	26: What does the action spectrum measure?
21: Which wavelengths does ozone protect us from most?	19: What does the strong absorption of UV radiation in the ozone layer do?

Original order	Practice order
22: How much would a 10% decrease in ozone increase the amount of DNA-damaging UV?	20: Why is it safer for living things at lower altitudes?
23: Why does time of day and season affect how much UV radiation is absorbed by ozone?	22: How much would a 10% decrease in ozone increase the amount of DNA-damaging UV?
24: What is solar flux?	23: Why does time of day and season affect how much UV radiation is absorbed by ozone?
25: What unit is solar flux measured in?	24: What is solar flux?
26: What does the action spectrum measure?	25: What unit is solar flux measured in?

How to display your questions

You can write your questions on cards if you like, and if you anticipate needing a lot of review to master this material, this may be a good option. However, most learners can probably settle for the much easier strategy of writing the questions down in one or (preferably) two lists (the second list is for your alternative questions), with the answers on the back of the sheet. You may be tempted to omit the answers, on the grounds that you have your set of learnable points to refer to, but I do recommend actually physically writing out the answers — not only because the process will help reinforce the answers, but also because you will find it easier to check your answers. It's easy to omit this step when you are sure of your knowledge, but, especially in the early reviews, it really does pay to check!

Points to remember

There is no one-to-one correspondence between learning points and questions — one learning point may generate several questions.

How much you break the learning point down depends on your background knowledge and understanding of the material.

Your questions should be neither too easy (if you already know the answer perfectly well, it's pointless to test it) nor too difficult (if you don't understand the material sufficiently well to answer the question, then you need to work on that more first).

Take care to frame questions so that they 'stand alone' and don't rely on your memory of the original sentence.

Include variants of your questions, to enable you to review with different questions.

Put your questions in an order that makes sense to you, and keep that order for the initial reviews, only mixing it up when you are confident of your understanding.

Concept maps

The second strategy for practicing complex text that I want to discuss is concept mapping. Earlier in the book, I mentioned research showing that concept maps, when used in the presence of the text, were no more effective than re-reading. I am actually a big fan of concept maps, so this shouldn't be taken as a slur. The point is how to use them effectively.

Concept maps are great for organizing your knowledge, and thus especially useful for understanding complex topics and forming those strong, meaningful connections. Concept maps are also a good strategy for 'priming' your mind — getting yourself into the right head-space for studying a topic, attending a lecture, doing a test.

Most importantly in this context, concept maps are also a good strategy to use with retrieval practice.

When I discussed them earlier I kept making the point that in the study the concept maps were drawn in the presence of the texts. Actual retrieval wasn't required. But if you *don't* have the text in front of you, if you draw a concept map as a means of retrieval practice, then this is a completely different story.

Concept maps are a good strategy for retrieval practice because:

- Many people find them more enjoyable than, say, writing down a list of points, or answering a list of questions.

- They can be a little different each time, giving you the opportunity to make new connections.

- They provide a spatial visualization — and spatial information tends to be more easily remembered.

Concept maps are not as widely known as they should be, not as well-known as mind maps are. I'm sure you're familiar with mind maps, even if you haven't used them yourself. Here are some of the important ways the two differ:

- In a concept map, lines connecting the concepts are labeled to show the relationships between concepts.

- In a mind map, the main themes are connected only to a single central image, not to each other, and the connections between concepts are not labeled.

- Mind maps are usually more visual, and concept maps more verbal.

The fact that the links are all labeled is a crucial factor in what makes concept maps such an effective tool for developing understanding, and why I prefer them over mind maps as a means of taking notes. It's one thing to realize that two concepts are connected; it's quite another to be able to articulate the nature of that connection. However, when it comes to retrieval practice, you may find the simpler mind map easier to use.

My own recommendation would be to use a concept map for initial review, when you're still consolidating your grasp of the material, and then perhaps using a mind map or simplified concept map (without labeling the links) for later reviews.

Drawing concept maps for understanding is something I cover in considerable depth in *Effective note-taking*, but for retrieval practice we can keep it simple. Here are the basic steps:

1. Articulate the question that is your main focus.

2. List the key concepts.

3. Describe the attributes of these concepts.

4. Articulate the relationships between the concepts.

5. Order the concepts in a rough hierarchy from most general to most specific, in this context.

Let's see that process in action:

1. Focal question: Why is ozone important?

2 & 3. Key concepts, with attributes:

- **ozone**: a type of oxygen molecule found in small concentrations in the atmosphere

- **ozone layer**: holds 90% of atmospheric ozone; in stratosphere; protects against UV-radiation; entirely blocks UV-c, partially blocks UV-b

- **tropospheric ozone**: holds 10% of atmospheric ozone; pollutant; large component of smog

- **wavelengths**: shorter wavelengths are more energetic than longer ones; very energetic wavelengths can damage DNA by breaking their bonds

- **visible spectrum**: wavelengths humans can see; range 400nm to 700nm; red at long end (630nm); violet at short end (410nm)

- **ultraviolet waves**: wavelengths shorter than violet; range 1-400nm; three types: UV-a, UV-b, UV-c (longest to shortest)

- **DNA**: the instructions in biological cells

- **solar flux**: a measure of solar energy; varies depending on time of day and season; measured in watts per square centimeter

- **action spectrum**: a measure of radiation damage

4. Relationships between concepts:

- **visible spectrum** contains a subset of **wavelengths**
- **ultraviolet waves** are beyond the **visible spectrum**
- **ultraviolet waves** can damage **DNA**
- **ozone layer** blocks most of the **ultraviolet waves**
- **ozone layer** contains 90% of atmospheric **ozone**
- **tropospheric ozone** is 10% of atmospheric **ozone**
- **solar flux** measures **ultraviolet radiation**
- **action spectrum** measures **ultraviolet radiation** damage to **DNA**

5. Concept order (most general to most specific):

- **ozone**
  - **ozone layer**
  - **tropospheric ozone**
- **wavelengths**
  - **visible spectrum**
  - **ultraviolet waves**
  - **solar flux**
- **DNA damage**
  - **action spectrum**

On the next page you can see an example of a concept map produced from this.

Do note that there's no 'right' answer! When using concept maps for retrieval practice, the important thing is simply to make sure that you get all the key concepts. You don't have to write down all the details — but do say them (preferably aloud) to yourself.

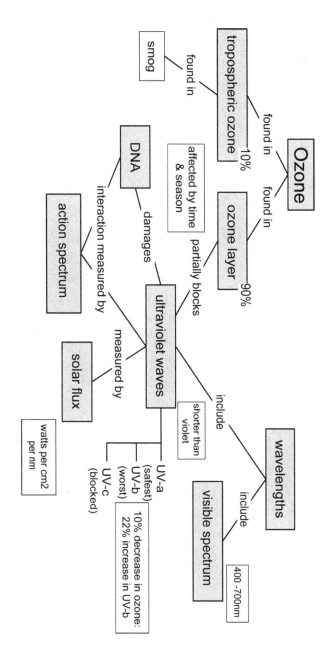

You can, of course, add as much detail as you wish to your concept maps. However, your review maps don't need to be as complete as those you make when mastering the material.

Do remember that you don't need to choose between questioning and concept maps for your review strategies. My own recommendation for revising complex texts would be to use both Q & A and concept maps, on different review occasions.

Points to remember

Flashcards (or flashcard software) are very good for learning foreign or technical words, and can also be used for simple facts (e.g., What is the capital of Australia?).

The keyword mnemonic is an effective strategy for making difficult words or facts more memorable, and can be used in conjunction with a flashcard strategy.

Complex information can be practiced by transforming the learnable points you have generated from the text into sets of questions.

Concept maps can also be used to practice your retrieval of these learnable points, and you can improve your learning by using both concept maps and questions on different review sessions.

Skill learning

> Learning a skill, whether a motor skill like playing golf or a musical instrument, or a cognitive skill like chess, is a different process than learning information.
> Accordingly, some changes must be made to our list of effective principles. In particular, while retrieval practice is the focus of effective information learning, deliberate practice is the central element in effective skill learning.

Until now, I've spoken as though learning information and learning skills is the same thing, not separating research using skills such as typing from research using text. To a large extent, the same fundamental principles apply to these very different domains, but they also differ in some crucial ways.

Before I start, let me make clear that skill learning is a large topic, one which could easily occupy an entire book by itself. I don't intend to give it the same in-depth coverage I've applied to information learning, which is the main focus of this book. I especially don't intend to discuss at any length skill learning as it applies to sport! Specific instruction in individual sports is covered by those expert in those specific areas, as it should be.

My aim in this chapter is rather to sketch out the ways skill learning differs from information learning, to discuss the role of deliberate practice in effective skill learning, and to give you some pointers that will, I hope, help you successfully approach cognitive skill learning in particular. I also try to clarify when you are learning a skill rather than information — a distinction that is not always as obvious as it may seem.

When you think of skill learning, you probably think of learning a sport, learning to play a musical instrument, to type, to ride a bike or drive a car. These are typical examples of skill learning, and the common element is that these are all motor skills. But there is another type of skill which is more relevant to study, and that is cognitive skills. A typical example of a cognitive skill is chess, but of course reading and study skills are also cognitive skills. Moreover, many study topics are not only informational, but also contain cognitive skills. Mathematics, chemistry, medicine, and foreign languages are prime examples of these.

We are all well aware that skills are learned and remembered differently than information. You probably think that motor skills and cognitive skills are equally different from each other. Well, that's true in a sense, but all three of these share similarities. Cognitive skills are more similar to information memory than motor skills are, but most motor tasks have an information component, to a greater or lesser extent (something which significantly affects some aspects of practice).

In other words, while these three domains have important differences, they can be thought of as all belonging to a continuum of memory and learning, with information at one end and physical actions at the other, rather than as completely separate domains. This is despite the fact that information and physical actions are encoded and processed quite differently.

In most of what follows, I focus largely on motor skills. Later in the chapter, I'll talk about cognitive skills specifically.

Skill learning begins with instruction

Skill learning is thought to go through three stages. It begins with instruction, which usually involves both verbal instruction and a physical demonstration of the skill. The quality of that instruction is one factor affecting how quickly you learn a skill.

We've all had experience of good instructions and bad instructions, even if we can't articulate exactly why the instructions are good or bad. In regard to physical skills, one factor has to do with the direction of focus. Instructions that tell you what to do with your body ("The right index finger and thumb continue to grasp the short end ...") are often not particularly effective. Instead, instructions are usually better to direct your attention to the *effects* of those movements. So, for example, it's often better for golfers to be told to focus on the swing of the club rather than the swing of their arms.

It seems that directing attention to an external rather than internal focus (the tool being manipulated or the thing in the environment being interacted with, rather than any part of your body) helps shorten the first stage of skill learning, speeding up the progression to a less conscious, automatic process.

Having said that, this advice depends on the specific skill being learned, and also on the degree of expertise you have. At certain stages of skill learning, instructions specifying precisely what you do with your body may well be more effective. The point to bear in mind, then, is simply that this distinction exists — that instructions may direct your attention to your body or to something outside your body, and if one type of instruction isn't working well for you, to try the other.

Explicit instruction, however, is not the only means of teaching a skill, and for motor skills verbal instruction is almost invariably accompanied by a physical demonstration of the skill.

Modeling

We now know that neurons in our brains activate in response to other people's actions in the same way (although at a lower level of activation) that they respond to our own actions. We experience this echo effect every day, when we wince or clutch at our body on seeing someone else being hurt, when we observe someone dancing and feel an answering impulse in our body, when we yawn on seeing another person yawn.

This imitative behavior is not simply the key to how children learn, but enables all of us to benefit from observation as a form of practice. That observation doesn't have to be of an expert.

One useful practice method for physical skills is collaboration in pairs. Research has found that, when paired individuals alternate between performing the skill themselves and observing the other, both individuals learn at least as much, and sometimes more, as they would if they'd spent the same amount of time practicing the skill.

There are two main advantages that paired practice might give. The first relates to the greater variability of the practice, since people are likely to perform differently, despite being instructed in the same way and provided with the same demonstrations. Consistent with this, paired practice has been found to encourage more flexible performance. The second advantage relates to motivation. Paired practice is often more enjoyable than solo practice, and also can provide the spur of competition.

Of course, if your partner makes you feel inadequate, or performs poorly, such practice may do more harm than good! As always, the general principle must bow to the specifics of a situation, and your own personality.

Video demonstrations, done well, can also provide excellent models. One of the big advantages of observation is that it provides an opportunity to focus on specific aspects in a way that might not be possible when you're performing the action yourself. A video, which you can stop and freeze and replay, as often as you like, is great for this.

Observing others, then, is a useful strategy that may help you short-cut your learning of physical skills. Just be aware that you can equally easily 'pick up bad habits' by observing poor performances!

Automating action sequences is the heart of motor skill learning

The initial instruction phase is followed by a period during which you learn to coordinate the various physical actions and strengthen the connections between successive actions. During this stage, you still need verbal reminders to tell you what to do. In the final stage, however, you lose the verbalization entirely.

This is the great difference between learning a skill or motor sequence, and learning information. Once the sequence has become automatic (through repetition, that is, practice), you can do it faster and without

Instruction
verbal instruction/physical demonstration

↓↓

Coordination
of action components & motor sequences

↓↓

Automatization
of actions & sequences

putting much load on your working memory (thus freeing up working memory for 'higher-level' strategic thinking). The downside is that, once it reaches this stage, if you do 'engage your brain' and start to actually think about what you're doing, your performance of it will almost certainly deteriorate. We're all familiar with this!

How 'muscle memory' is different from information memory

I said that we all know that learning and remembering skills is different from learning and remembering information. This is reflected in the way we talk about 'muscle memory'. Despite its name, the memory is held in the brain rather than in the muscles themselves (although of course the muscles will also be changed by the practice of the movement). But the term 'muscle memory' isn't, I don't think, so much a sign that people really believe memory is held in the muscles, as it is simply an acknowledgement that this type of memory is noticeably different from other types of memory. In particular, it seems to be far more durable, more resistant to the effects of disuse. There are several reasons for this.

Motor skill learning takes place in a different region of the brain than information learning (known more formally as **declarative memory**), and this region is in one of the oldest parts of the brain (the **hindbrain**). Motor learning transfers from short-term to long-term storage much more quickly than declarative memories: within a few hours in some cases, within a few days at most (in contrast to the weeks before declarative memories are completely transferred).

This difference in timing has important consequences for effective practice.

Additionally, the transfer process tends to drop what it regards as unimportant details to an even greater extent than occurs with declarative memory. In other words, only a small part of the information you have encoded in your short-term memory will be passed on to your long-term memory. Thus, while sophisticated subtleties of motion and action are easily noted and encoded in short-term memory (enabling, for example, a baseball batter to adjust to the pitches), such subtleties are usually lost — within some 10-15 minutes, once the event is over. Quickly in and quickly out, seems to be the motto!

What this means is that only the broad outlines of a procedure get locked into long-term memory. Muscle memories are easily acquired and the details are just as readily lost. On the plus side, they are also quickly picked up again with a little practice. All this is reflected in the way we talk of motor skills getting 'rusty', our knowledge that we have to constantly use skills to keep them honed, and our confidence that, if we have neglected them, nonetheless we will get them back with a little work. It's also why even experts don't just walk into a situation of using their skills (a concert; a ball game) without some warm-up.

Because muscle memory is so strong once properly consolidated, and perhaps also because distinguishing details are likely to be lost, interference is even more of a potential problem than it is for declarative memory. This means you need to watch out for steps within the new skill that are antagonistic to steps contained in an already mastered skill — for example, if you decide to move from using a standard QWERTY keyboard to a Dvorak one (where the letters are arranged differently), this is much more difficult than if you start off with a Dvorak keyboard from the beginning, and your difficulties will be directly linked to how well you typed on the QWERTY keyboard.

As a general principle, the better your learning of the first skill, the harder it will be for you to learn an antagonistic skill. However, if a new skill is *compatible* with an old skill — if it has action sequences in common — then the better you are at the old skill, the faster you will learn the new.

Because skills are encoded differently than information, because they are so quickly consolidated yet so vulnerable to interference, the way you practice them is even more crucial to your success. In particular, spacing and interleaving are even more important, and so is the content and quality of your focus.

Points to remember

The key factor distinguishing skill learning from information learning is its potential for automatization, which is only achieved through practice.

Motor skills, perceptual learning, and information learning are all processed in different brain regions, and processed differently.

The 'gist' of actions is transferred very quickly to long-term memory, and remembered for a very long time even if not used. However, details are quickly lost, meaning that, if you want to keep a motor skill 'sharp', you need to keep using it.

Learning a new skill will be helped or hindered by the extent to which it shares elements with existing skills, or contains antagonistic elements.

Spacing and interleaving are critical factors in effective skill learning.

Deliberate practice

All of this is tied up with the concept of 'deliberate practice' (a term coined by K. Anders Ericsson, the guru of research into expertise). Ericsson makes a very convincing case for the absolutely critical importance of this type of practice, and the minimal role of what is commonly termed 'talent' (you will no doubt have heard of his '10,000 hours' theory, which was popularized by Malcolm Gladwell). His research shows that experts only achieve their expertise after several years (typically ten or more) of maintaining high levels of regular deliberate practice. But most people, he suggests, spend very little (if any) time engaging in deliberate practice even in those areas in which they wish to achieve some level of expertise.

So what distinguishes deliberate practice from less productive practice?

Ericsson suggests six factors are of importance in deliberate practice:

- The acquisition of expert performance needs to be broken down into a sequence of attainable training tasks.

- Each of these tasks requires a well-defined goal.

- Feedback for each step must be provided.

- Repetition is needed — but that repetition is not simple; rather the student should be provided with opportunities that gradually refine his performance.

- Attention is absolutely necessary — it is not enough to simply mechanically 'go through the motions'.

- The aspiring expert must constantly and attentively monitor her progress, adjusting and correcting her performance as required.

Breaking down a skill

Knowing where to divide a skill is something that develops with practice. For a musician, it might mean practicing a piece in appropriate sections that are only a few bars long (maybe only one!), with special focus on those that are difficult. For a sportsperson, it means breaking actions into self-contained components with a special focus on one specific aspect of performance.

For example, one long-time professional golfer (Moe Norman), renowned for being a great ball-striker, attributes his success to the systematic practice he began at 16. For well over ten years, he hit 800 balls a day five days a week (on the weekends he played golf). During this practice, he always had a specific focus, such as hitting to a bucket, hitting to an image of a specific hole, controlling a particular aspect of his swing. As a result, he swings at the ball with incredible consistency, always straight, the ball always going precisely where he wants it to go.

This focus on segments (musical sections; action components) harks back to the 'chunks' I mentioned in regard to working memory and information learning. Regardless of whether you're trying to master an academic subject or a physical skill or even a visual task (such as learning to recognize the works of different artists), you always need to break it down into manageable chunks (which may be very small at the beginning, because working memory is very small), and you always need to be able to judge (through feedback and self-monitoring) the relative difficulty of those chunks, so that you can apportion your time and effort appropriately.

'Appropriately' means spending more time and effort on difficult bits and less time on easy ones. This may seem obvious, but many a student has done it the other way around, on the grounds that one is hard and the other is easy! That's all very well if your aim is simply to 'get through' a specified practice/study time, without any wish to achieve much, but if you do actually want to improve, you need to focus on the hard stuff.

But you don't get points for making it more difficult! If something is too hard for you, keep breaking it down until you have a piece you can handle. Remember that working memory is very small, and the key to increasing its capacity is to increase the size of your chunks. You do that bit by bit, not by trying to leap to where you want to be. It is not, therefore, an admission of weakness or failure to begin with very small chunks (sections, elements, movements); the size of the chunk merely reflects your previous relevant experience. Appropriately directed experience will grow your chunks.

Varied repetition

As Ericsson says, while repetition is vital, that repetition shouldn't be 'simple'. To refine your performance, you need variation. You can't work out the best way of performing an action without trying out subtle variations, such as varying the force of your pitch, the distance you are throwing, the angle of your arm, etc. But 'refining' your performance is about more than that — it's about making it more flexible.

A distinguishing difference between novices and experts is that novices are less adaptive. They find it harder to adjust their performance in response to changes in conditions (such as weather changes, for drivers; a different acoustic environment, for musicians; a change in pitch, for cricketers). This isn't surprising — experts develop abstract ideas of what they're doing; they know what's 'core' and what (and how) other details might change. Novices are still performing by rote. To develop the necessary abstraction, a mental **schema**, you need to experience variations.

This is similar to the way we develop concepts in semantic memory: we build the concept of 'dog' by experiencing a variety of dogs and non-dogs (such as cats), working out what features the dogs share, how they're different from non-dogs. It would be very hard to build a good 'dog' concept if you only knew corgis!

In a similar way, as you build a skill, you build an abstract idea of the skill — which features are important, which are not, what goes together and how, what leads to what. To learn all this, to build the abstract schema that will guide your performance across a range of contexts, you need to vary your experiences and actions.

Variation isn't, then, solely a matter of subtle changes in your movements, such as changing the angle of your arm or the force of your throw. It's also about context. Thus, a study in which baseball players learned three different kinds of pitches found that those who practiced the pitches in separate blocks learned significantly more poorly than those who practiced them in blocks in which different pitches occurred randomly, with no one type of pitch occurring more than twice in a row. Those in the random (interleaved) group performed better when tested at the end of training, and showed more than twice as much improvement from training (57% vs 25%). (Do note, however, that blocked practice was still far better than less practice! A control group, that only experienced 12 'acquisition' sessions that were the same for all groups, and not the extra 12 sessions given to the random and blocked groups, only improved by 6%).

Interleaving is therefore of particular value in skill learning. There are two aspects to this. In the previous example, interleaving different pitches mimics what happens in a real game, providing the opportunity to practice the necessary skill

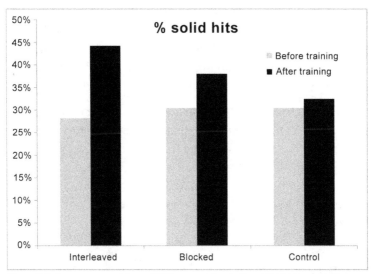

Figure 9.1: Average percentage solid hits, comparing performance before and after baseball training, for interleaved, blocked, and control conditions. From Hall et al. 1994.

of recognizing the type of pitch and responding appropriately. This is hardly specific to baseball pitches! In almost all real-world scenarios, we're required to make such judgments and adapt our performance accordingly.

The second aspect is that benefit of interleaving that I discussed earlier: the way it helps you see vital similarities and differences. If your new skill contains components (action sequences, movements, musical sections, etc) that are similar but not identical, interleaving them, although it will slow your initial learning, will pay off in the long run.

Do remember that interleaving involves *repeated* mixing of tasks! Simply following one task (set, sequence, skill) with another is very bad for learning, since it promotes interference without any mitigating factors. Thus, in one study, music students learned to play a short melody (a 13-note sequence) on the piano in the evening, before being tested on it the following morning. Some students learned only one melody, while others learned this melody and then another. A further group also learned both melodies, but briefly practiced the first melody again at the end of the training session. Only the first melody was tested in the morning.

Those who learned only one melody showed the usual post-sleep consolidation jump that I discussed earlier. Those who learned another melody after the first one showed no such improvement, demonstrating how such interference blocks consolidation. But those who played the first melody again, after training on the second melody, did show the usual post-sleep improvement.

Feedback

Feedback has two main roles in practice: monitoring, and motivation. Let's talk about motivation first.

Do I need to say that motivation is important? Sometimes we dismiss motivation as if it's a luxury, a non-vital factor that only comes into play when the individual lacks the self-discipline to do what they know they should do. But 'motivation' has to do with having a 'motive', and a 'motive force' is a force that makes things move. It is motives that drive our actions, and motivation is vital to actions and to practice. It's not simply about giving someone money, or a treat, or a gold star.

Motivation is about having a goal, and being able to maintain that goal.

Feedback is a factor in providing motivation because it can help you stay committed to your goal. It can also cause you to discard your goal, to give up. It's crucial, therefore, to get feedback right.

In the context of motor skill learning, feedback is generally concerned either with the outcome of the action or the quality of the action. It may be explicit (such as your coach or teacher telling you exactly what you did wrong) or implicit, inherent in the action itself (if you're aiming at a target, or if you're trying to sink a putt, it's obvious how well you've done).

Implicit feedback is less potentially dangerous than explicit feedback. You may decide to give up because you're not doing as well as you think you should, but that's a problem of your expectations and goal-setting, rather than a problem inherent in the feedback. Implicit feedback is vital for good self-monitoring. Explicit feedback, on the other hand, is more problematic. When we think of feedback as a de-motivator, it's generally external feedback that's the problem.

Research indicates that, in the context of skill learning, negative feedback — feedback about errors, or feedback telling you that you're not doing as well as your fellow students — can worsen performance. However, feedback telling you that you're doing well (regardless of its accuracy), improves performance.

It's suggested that one reason why negative feedback might hamper skill learning is that it increases thoughts of 'self'. This comes back to the third phase of skill learning — the aim is to remove consciousness from the process. When you're thinking about your self, when you're fretting over your performance and thinking about exactly what your body is doing, you're standing in the way of the proper performance of the skill.

If your coach, teacher, or fellow-students are overly critical, therefore, you might find it helpful to either:

- change them for more supportive people!
- try to convince them that negative feedback is counterproductive, or

- find some occasions to practice without them.

But of course feedback has another, entirely positive (indeed crucial) role: monitoring.

Accurate self-monitoring is critical to successful practice. You can't do it right if you don't know when you're doing it wrong! While coaches and teachers (and more experienced peers) can certainly provide explicit feedback that helps you monitor how you're doing, it's still a good idea to learn to do it yourself even when you do have such resources. Sure, when you start with a new skill, some sort of guide is extremely helpful, if not critical. However, once you've grasped the basics of the skill, your aim should be to regard mentors as additional resources rather than primary. No one can be at your shoulder at every minute of your practice, and in any case, learning is always individual— only you know how your body feels on the inside, what your mind thinks and feels.

This is probably why external feedback seems to be most effective when it's only given on some occasions, rather than all. If you know you're going to get feedback every time you do something, you have less incentive to pay attention to what you're doing, less incentive to learn how to assess and monitor your own performance.

Timing is another issue with explicit feedback. Implicit feedback occurs naturally in the course of performance, but explicit feedback, of course, can be given at any time. Previously, in the context of information learning, I said that feedback can be effective even when a little delayed. However, with motor skills, because of the speed with which muscle memories are consolidated, it's more important that feedback occurs quickly. Another opportunity for explicit feedback to get things wrong!

Self-monitoring and goal-setting

The point of monitoring is to help you work out when you're doing things right and when you're doing them wrong. That's not simply about recognizing errors, although error recognition is an important part of self-monitoring. You can, after all, want to improve your performance from "not bad" to "good", or from "good" to "better". Self-monitoring — watching and assessing your performance — helps you adjust and adapt your performance appropriately.

So how do you improve your monitoring skills?

One factor that helps considerably is explicitly articulating your immediate goal. Science, it's said, has to be testable; if you can't test the theory, it's not scientific. In the same way, a goal isn't really a goal if you can't tell when you've met it. A clearly specified goal should give you the information you need to assess your performance (and by 'should', I mean that, if it doesn't give this information, you need to specify it better!).

A vital part of deliberate practice, therefore (as it is with all learning!), is to clearly articulate your goal.

A useful distinction here is that between process and outcome goals. Process goals relate to fundamental techniques, to mastering specific procedures. Outcome goals relate to the outcome or product. So, for example, in learning to throw a dart, the outcome goal might be to hit the bull's eye or to achieve a specific score, but the process goals might relate to the proper execution of a specific forearm throwing motion, or extension of the fingers.

A study in which 90 school-girls were taught to throw darts compared several types of goals:

- no set goal

- outcome goal ("try to attain the highest possible score")

- process goal ("concentrate on properly executing the final two steps in every throw")

- transformed goal (depending on where the dart strikes, concentrate on one of these steps on the next attempt)

- shifting goal (students were given the process goal, before shifting to the outcome goal after 12 minutes of practice).

Those given the shifting goal learned best and became most interested. The worst (excluding those not given any goals at all) were those who were only given outcome goals. The graph below shows dart-throwing skill measured by the average of six throws at a target made of seven concentric circles, with each circle ranging from 1 (outermost circle) to 7 (center). The study also included a further condition: some of each goal group recorded their performance as they practiced, either in terms of their scores or their actions, as appropriate. As you can see, whatever the goal, this self-monitoring significantly helped.

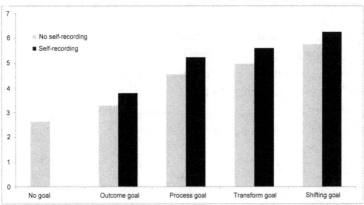

Figure 9.2: Average dart-throwing score (out of 7), comparing goal condition and self-monitoring during practice. From Zimmerman & Kitsantis, 1997.

It's worth noting that one of the advantages of process goals seems to be that it encourages the learner to attribute any failures to strategy rather than lack of ability. In this study, the girls who were given process goals were more likely to attribute their deficient performance to a strategy than were those girls given an outcome goal or no goal. This in turn led to having a greater belief in their ability to learn, and more enjoyment in the task.

Self-monitoring and goal-setting are both examples of what is called **metacognition**, which includes both your understanding of cognitive processing in general and your understanding of your own cognitive processing. Let's look at how metacognition helps skill learning, using the example of a motor skill with a high cognitive component: learning to play a musical instrument.

Metacognition and self-monitoring

Part of the cognitive component in learning to play a musical instrument is theoretical knowledge (e.g., what's the C major scale) and, less obvious but more vital, aural schema (mental patterns of how different genres of music 'should' sound). Another part is concerned with specific pieces of music. While playing a specific piece is very much a skill, in that it's a specific motor sequence, it's not *only* a sequence of physical actions. There's also processing that encodes visuospatial information (such as the sight of particular piano keys going down in a particular order), encoding of the temporal rhythm (this isn't only a musical phenomenon — rhythm is an important part of many motor skills), and of course sound information.

There have been several studies exploring how practice differs between experienced musicians and beginners. What's been found is that, while there are certain tendencies in strategies used, the fundamental difference between experts and novices lies in metacognition. Experts understand their own strengths and weaknesses, understand the different tasks involved in playing music, recognize when approaching new music which parts of it will be more difficult for them. Experts, you could say, approach learning strategically.

This applies across the board; it's not specific to musicians. The heart of successful learning, and building expertise, lies in your metacognitive abilities. This is why, although expertise is specific to a domain or topic, if you're an expert in one thing, you have a head start in mastering another subject. You know how to be an expert; you know how an expert approaches learning.

Consistent with this, almost all experienced musicians get an overview of a new piece first, either by playing it through or examining the score. All of them stress the importance of thinking analytically about the music and playing it very slowly and carefully in the initial stages of learning it. After this, however, there seems to be considerable variation, in, for example, the level of organization and planning, and the extent to which they mark the score.

Another consistent factor in developing expertise lies in the response to errors — also a metacognitive act. Novices tend not to recognize when they have made errors, and when they do, often ignore them (not necessarily deliberately). The vast majority of novices practice by repeatedly playing a piece straight through. One of the problems with this is that they are far less likely to remember any errors they made, and when they repeat the piece, the chances are that they'll repeat the errors too, and cement them into the music.

Here's a list of strategies associated with growing expertise among students (beginner to Grade 8). Notice which ones have to do with pinpointing difficult sections or aspects and dealing with them deliberately:

- Practicing small sections
- Getting recordings of a piece that is being learned
- Practicing things slowly
- Knowing when a mistake has been made
- When making a mistake, practicing a section slowly
- When something is difficult, playing it over and over again
- Marking the score
- Practicing with a metronome
- Recording practice and listening to tapes
- Identifying difficult sections
- Thinking about how to interpret the music
- Doing warm-up exercises
- Starting practice with studies
- Starting practice with scales.

By contrast, the following strategies were negatively associated with growing expertise (that is, they became steadily less common as the students became more expert):

- Practicing pieces from beginning to end without stopping
- Going back to the beginning after a mistake
- Immediately correcting errors (that is, 'fixing' a mistake by stopping and immediately trying to play the note

correctly, as opposed to stopping at the end of the section, then playing the section again more slowly and with more attention).

It's worth noting that these strategies might well be appropriate right at the beginning, when the student is first learning to play the instrument, but they become less useful as pieces become longer and more complex. It's also likely that at higher levels of expertise (say, among professional musicians), some of the 'good' strategies may be less appropriate. This is another reminder that the most effective strategies often vary depending on your level of experience and knowledge.

Similarly, although you can short-cut your path to expertise by using metacognitive strategies, you can't avoid the need to first build up a certain level of basic skill and music knowledge. One reason for that is that playing music, whatever the instrument, requires several concurrent processes. For example, for a violin, there are the finger movements, the movement of the bow, reading the notes, attention to pitch, to rhythm, to intonation, to dynamics. You can't master all of these all at once, and trying to do so will only hurt your learning.

The successful learner changes strategies over time and with experience. With a motor skill, you need a basic foundation in place before getting 'strategic', but early adoption of a metacognitive approach — one in which you are attentive to your performance and aware of different strategies for improving it — will speed your progress considerably. Remember that, while no specific strategies can be unequivocally pointed to as critical, the adoption of *systematic* practice strategies is significantly associated with expertise (accounting, in one study, for around 11% of the difference between individuals, which is more than any other single factor).

In other words, finding effective strategies is important, but effective strategies are specific to the individual. You can't assume that because something is the best strategy for your best friend, that it's going to work equally well for you. What's important is that you find *your* effective strategies, and practice them regularly (that's the 'systematic' aspect).

How do you develop metacognitive skill? One part of it is what you're doing now — learning about learning and how your brain works; learning about different strategies that may be effective for your specific task. The second part is learning to monitor.

I said before that one way to improve your monitoring skills is to get in the habit of explicitly articulating your immediate goal. Another helpful strategy is explicitly verbalizing what you're doing and why. You don't have to do this every time you perform! But if you make the point of doing it periodically, you'll develop your monitoring skills, and also find it very useful for your specific practice.

Note that you don't have to do it either completely or not at all — having verbalized the whole action sequence, you should have some idea what the difficult elements are; you can therefore make a point of verbalizing at those points only.

You also should consciously think about your performance and your learning process — not in a "oh it was so awful" way, but in an analytical "mistakes are there to learn from" way — and discuss it with others (learning to listen to others' criticism, and to give helpful criticism to others, is itself a skill that gets better with practice!).

Self-monitoring can be an annoyance if you're new to it — you may be inclined to see it just as something that slows down your practice. But keep in mind that monitoring is a skill, and will get faster and less hampering as you become more practiced at it. It really is worth forcing yourself through the initial pains, because self-monitoring is a crucial skill if you're serious about practicing and learning effectively.

The importance of self-monitoring is part of the reason why the core attribute of deliberate practice is constant and intense focus. This need for a high level of concentration is why deliberate practice is limited in duration. Whatever the particular field of endeavor, there's a remarkable consistency in the habits of elite performers that suggests 4 to 5 hours of deliberate practice per day is the maximum that can be maintained.

This, of course, can't all be done at one time without resting! As I discussed earlier, you may find it helpful to concentrate your activity in short 15-minute bursts, followed by brief periods of rest. Five minutes 'quiet time', during which the skill will consolidate in your brain, is worth a great deal, and, as I mentioned, is better earlier in practice than later. This may be of greater benefit for novices — as you gain experience, you should be able to increase those activity periods. Even for experts, however, an hour seems to be long enough, and day-time napping is common among elite performers operating at these high levels of concentration.

Not all practice is, or should be, deliberate practice

Deliberate practice is hard work. While it's critical for improvement, those engaging in physical pursuits also need other activities that are aimed at general fitness. Additionally, there may be tactical or theoretical issues to master. And, of course, the role of 'fun' activities shouldn't be overlooked either. You aren't going to persevere with a skill if you never enjoy exercising it (going back to the importance of motivation).

In general, experts reduce the amount of time they spend on deliberate practice as they get older. It seems that, once a certain level of expertise has been achieved, it isn't necessary to force yourself to continue the practice at the same level in order to maintain your skill. However, as long as you wish to improve, a high level of deliberate practice is required.

But you shouldn't take the '10,000 hours' too literally. It's now clear that there is considerable variability between individuals. While Ericsson was certainly right, and groundbreaking, when he propounded the virtues of practice and a diminished value for 'natural talent', talent is still a factor. There's no doubt that some people need far less practice to achieve the same (or higher) levels of performance compared to others. Having said that, it's not yet clear to what extent this is a matter of individual talent, or simply a product of using the practice time better. That is, for whatever reason (natural ability; a wise mentor; dumb luck), some people practice effective strategies from an early stage.

None of this is to assume that your aim is, or should be, to become expert at whatever skill you're learning! I discuss how experts differ from novices in order to show the path; you can stop at any point on the path to expertise. We learn, or want to learn, many skills, and all of these at different levels of expertise. How skilled you want to become is part of your outcome goal.

Points to remember

A necessary, though not sufficient, requirement for effective skill learning is regular deliberate practice.

Deliberate practice is focused and goal-oriented. In deliberate practice, you select a manageable chunk, specify a goal, monitor your performance and adjust it accordingly.

Goals should be responsive to your performance and needs, aimed either at specific steps/techniques, or at outcomes, as appropriate.

Variation, in your performance and in the environment, helps you develop good mental schemata that enable you to adapt to changing demands.

Deliberate practice is intensive and thus you're limited in how much you can do (effectively) at one time and in one day.

Motor skill learning can benefit from observing others perform, not simply as a demonstration of how to do something, but as a form of practice.

Feedback is most useful when it's an integral part of the action. The best explicit feedback is occasional and positive.

Mental practice

Another type of practice I should touch on is 'mental practice', that is, of imagining yourself practicing a skill. This idea has become popular, and deservedly so in some domains — but not perhaps in all those domains in which it has become popular! While it seems clear from the research that mental practice is of significant benefit to cognitive skills, its value to motor skills is not so certain. It's likely that a motor skill only benefits from mental rehearsal to the extent that it has cognitive elements. So, for example, it's useful to mentally practice solving a Rubik's cube, but less useful to imagine yourself typing.

The benefits of mental practice also depend on your imagery abilities. But don't rule yourself out if your visualization skills are poor! While the ability to visualize does vary markedly between individuals, the skill does respond to practice.

Given that a skill is high in cognitive elements, will mental practice always be helpful? Research indicates that it is most effective when you already have some experience with the skill. For beginners who are unclear about what they should be doing and how they should be doing it, it's only too easy for them to practice the wrong thing — for mental practice doesn't, of course, allow for much in the way of implicit feedback, to tell you how accurately you're performing.

Research also suggests that mental practice is of particular benefit when the task is complex — that is, when it contains many elements, when it puts high demands on working memory.

However, there is one benefit from mental practice that probably occurs across all types of skills, and for both newbies and experts: mental rehearsal for priming. By engaging in some mental rehearsal before performing the skill, you prepare the mind and body for the task ahead, helping you direct your attention to the aspects that matter.

Getting into the right frame of mind is something all sportspeople know about. It applies to other domains too, and it's not simply a matter of emotional state. It's about activating the parts of your brain that are involved in this skill. In music, this might be achieved by listening to a recording of a piece before starting to work on it. In study, by reading the Table of Contents, advance organizers and summaries (in the case of textbooks), or doing any required reading before a lecture, and (in both cases) thinking about what you expect to learn from the book or lecture.

Apart from working on your ability to mentally visualize objects and actions, you might improve your mental practice skills by practicing meditation-type exercises. That is, breathing exercises and exercises in which you focus on one specific part of your body or internal sensation.

As for the specifics of how to mentally practice your skill, here's an example of a training program aimed specifically at pianists, that will give you some idea:

- choose a small section of a piece you're learning (say, 4 bars)

- as you read the score, visualize as precisely as possible the keys on the keyboard corresponding to the written notes

- now visualize the position of your hand, the width of movement of your arm

- bring in the auditory 'images' — that is, hear the notes in your head as you mentally play them

- break down the section into its main components

- feel each single interval, in terms of both movement and sound, starting at a slow tempo

- feel inside your body how the fingers should press the keys, initially using a legato touch

- progressively increase the speed for each component

- occasionally try physically playing the passage, swapping back and forth between mental and physical practice

- near the end of your practice, abandon the legato touch and mentally play fortissimo

- play the whole sequence as a complete movement.

(adapted from Bernardi, 2013, who adapted it from Klöppel's 2006 mental training manual)

You can see in this description the process of breaking the task down that I talked about as so critical for deliberate practice. One step, however, is specific to mental practice — the recommendation to swap between mental and physical practice. I think this is probably an important strategy in building your visualization skills, and that the timing of it may also be important — not right at the beginning, and not leaving it too late. Finding the right time is, like so many other aspects of successful learning, a matter of becoming attuned to your abilities and performance.

Cognitive skills

I said earlier that cognitive and motor skills represent two ends of a continuum. The main factor determining where a skill is placed on that continuum is information content. To the extent that information underlies the skill, that skill is 'cognitive'.

Wrestling requires little in the way of knowledge; playing a musical instrument involves knowledge of music pieces and music theory, although its focus is on the physical action of playing; playing chess makes no significant demands on the body. Thus, wrestling is a motor skill, playing an instrument is a motor skill with a significant cognitive component, and playing chess is a cognitive skill.

Research has found that the most important predictor of chess rating is cumulative serious practice, of which the most important element is practice alone — that is, analysis of positions from chess books and other written resources. Note the power of serious study alone over other activities such as tournament play and analysis with others. Researchers suggest that this activity allows the most control over what is being practiced and for how long (it also gives the player the freedom to safely try out approaches they wouldn't risk in play with others).

What this deliberate practice of chess positions and movements does is grow a databank of patterns in the mind of a chess master. Research has shown that chess masters can look at a chess board in the middle of a game and easily remember exactly where each piece is (an ability which helps a great deal when performing the amazing feat of playing many games simultaneously!) — but this skill deserts them as soon as the pieces are randomly scattered about the board. Their impressive memory derives solely from the stored patterns, the chunks, in their long-term memory. Expert netball, basketball, and field hockey players similarly have been shown to have stored defensive player patterns that give them not only a better memory for play in their own sport, but in other sports as well.

To a large extent you can say that, while motor skills revolve around learning action sequences, the heart of cognitive skills (or the cognitive component of motor skills) is learning patterns. Learning the patterns important for your cognitive skill is essentially a type of perceptual learning. In reading, for example, some of the patterns are letter-combinations and words and their associated sounds and meaning, but the more sophisticated patterns that underlie reading *skill* are the patterns of words and phrases that allow you to anticipate what comes next. Indeed, prediction is central to skill, because it is the ability to anticipate what comes next, to prepare for the next neural activations, that enables the faster processing that is the hallmark of skilled performance.

In mathematics, patterns can be seen in the format of different types of problems ("The profit in dollars for the manufacture and sale of x soft toys is given by $P(x) = 50x - 0.002x^2$; Find the number of soft toys to be sold to maximize profits." "Find the derivative of $y = \text{cosec } x$"). In medicine, it's the recognition of certain symptoms ("I've seen that rash before; it looks like measles"), and the recognition of the grouping of certain symptoms ("Wheezes and crackles in his lower lungs; no fever; white blood count stable — I think it's chemical pneumonitis").

While cognitive skills require just as much practice as motor skills, that practice is aimed at building and recognizing patterns rather than automating action sequences. To recognize a pattern, you must extract from the mass of real-world data the bits that matter. This is not something that's readily taught. You can (and should) be guided (by a coach or mentor, or other resources), but at the end of the day, what's critical is varied experience.

Having said that, a guide can substantially reduce the amount of experience you need. While motor skills generally require a coach or teacher to provide that guide, for many cognitive skills this role can be taken by written resources. In chess that may be books analyzing chess movements and positions; in other cognitive skills, other types of models, such as worked examples, can be useful.

Worked examples provide models for cognitive skills

Worked examples — the cognitive equivalent of physical demonstrations — are particularly common, and valuable, in mathematics. However, they're not necessarily of equal value to everyone, or in all circumstances. Worked examples seem to be most useful to novices. Unsurprising, you may think, but the important point is that worked examples give the most help when you don't have a schema, a mental model, for the task. Once you have your own schema, worked examples can actually hurt your learning if they're not worked out in a way consistent with your schema.

In other words, not all worked examples will work for *you*.

One benefit of worked examples is to help reduce cognitive load, and for that reason it seems likely that they offer most benefit to students with a low working memory capacity. This is consistent with their greater benefit for novices — working memory capacity is much less of an issue for experts, who have schemata that reduce the load on their working memory; strategies to off-load some of the information onto other people or artifacts; the hard-learned skill of knowing what to pay attention to and what to ignore.

Like any model, the effectiveness of worked examples does depend on their design. Consistent with the fact that a principal benefit is to reduce the load on working memory, effective examples need to:

- avoid splitting the student's attention (as happens when

text and images are physically separated)

- avoid redundancy
- make subgoals (if any) explicit through labeling or by visually isolating sets of steps.

In other words, worked examples should be focused and explicit. If you don't find an example useful, it's probably because:

(a) there's more information than you need,

(b) it's missing information you need, or

(c) it puts too much load on your working memory by separating related information.

If (a), you can go through the example and draw a line through information that seems irrelevant. If (b), you should try and work out exactly what information is missing (usually steps that are obvious to the instructor), at which point you can seek help (finding an answer is much easier when you have a specific question!). If (c), you can try re-working the example so that related information is closer together.

Research also indicates that more experienced students can get the most out of worked examples if they explain each step to themselves as they go. However, beginners should be wary of this, as it's only helpful if they're capable of providing good explanations. Similarly, students with prior knowledge can benefit from seeing both correct and incorrect solutions (that is, good and bad examples), but newbies are better off seeing only correct solutions.

Although I've talked about when worked examples may not work well, that's merely by way of warning. Worked examples are valuable tools when it comes to learning cognitive skills. In mathematics, simple practice of modeled problems has been shown to be the highest predictor of performance on the type of problem modeled, even in the absence of feedback.

Automatization is the core attribute of all skills

Like motor skills, cognitive skills also become 'automated' in the sense that, with practice, many subprocesses no longer need conscious attention. This has the effect of reducing the load on working memory, and also reducing the number of steps that the expert is aware of — indeed, in familiar scenarios they'll often just 'know' the answer, and are at a loss to explain how they know. This explains why experts are often not the best teachers!

Similarly, the pattern recognition that is central to expertise in cognitive skills is a perceptual process that has become automatic — like instantly recognizing that a shoe is a shoe, or that *that* face belongs to Albert Einstein. Perceptual learning, which is encoded and processed in different brain regions compared to both motor sequences and information, is also, like motor skills, remarkably durable. One study, for example, found that recognition of complex stimuli, practiced over two days, were still remembered a year later.

Reading is a cognitive skill that displays both these attributes very clearly.

Reading is a process that encompasses a whole hierarchy of subprocesses, including the recognition of individual letters, the association of letters and letter combinations with particular sounds, the recognition of whole words, the knowledge of words and their meanings (semantic memory). A skilled reader has automatized all of these subprocesses, and can concentrate on the meaning of the text. The recognition of letters and words is, of course, a matter of pattern recognition.

Like reading, learning another language is a prime example of a study topic that combines both informational and skill learning. The information component is self-evident: all those new words and grammar points. The skill component is perhaps less obvious, depending on your (or your teacher's) approach.

Today, there's far more emphasis on conversation (nothing wrong with that! 'Practice the task you need to do.') But back when I learned Japanese at university, the prevailing theory encouraged something called pattern drills. These seemed to have fallen somewhat into disrepute nowadays, which I think is a shame, because they seem to me to be great examples of deliberate practice. The idea was that you'd be given a 'pattern sentence', such as this:

Sore o misete kudasai. (Please show me that one)

The pattern lies in only some elements: in this case, the *te* at the end of *misete* (show) and *kudasai* (please). In the drills, you would practice substituting different verbs and objects into the pattern. Thus:

Kore o tabete kudasai. (Please eat this)

Chotto matte kudasai. (Please wait a moment)

Language is made up of patterns. Think of how often certain phrases, even whole sentences and more, trip from your tongue, without any need for you to think about what you're going to say. Think of how often you 'know' what someone's going to say before they do. Knowing the patterns of a language, practicing them until they can be produced automatically, is a fundamental element in fluency.

Fluency, in other words, reflects automatization, and is therefore a skill, and can only be achieved through practice.

Approach skill learning like an expert

I talked before about the path to expertise, which is simply another way of saying 'skill'. You don't have to become an expert to learn a skill, but becoming skilled is what it means to become an expert. So the journey is the same; it's just a matter of where you stop.

I also talked about how an expert approaches learning.

Wherever you plan to stop, that is, however skilled you want to become, you can benefit from approaching skill learning as an expert does. What does that mean?

First of all, you need to understand yourself — how you think, what you're good at, what you're poor at, what things are easy for you and what things are difficult. Remember that the point of this is not to avoid things that are difficult, but to help you recognize what's difficult for you so that you can focus on it.

You need to understand the whole before getting bogged down in details. Spend time observing; don't rush in to try things out yourself. Get the big picture first.

Work out the different tasks involved in the skill, so that you can practice individual tasks separately before putting them together.

Consider different strategies for performing the tasks.

Motor skills and cognitive skills share the fundamental attributes of skills: goal-directed actions that achieve speed and fluency through automatization. But cognitive skills in particular benefit from taking a metacognitive approach — thinking like an expert.

Points to remember

Mental practice is useful for cognitive skills, or for the cognitive elements in motor skills, but less so for motor skills themselves.

Mental practice is particularly useful for complex tasks, and for practitioners with some experience.

How you approach learning is more important than any specific strategy for working on your skill — you need to think analytically about your performance, monitor it, and respond constructively to errors.

Deliberate practice of cognitive skills is focused on learning patterns, and automatizing subprocesses.

Because pattern recognition requires you to extract what matters from a mass of data, repeated experience is crucial, and a guide of some kind is extremely helpful.

Worked examples are most helpful when focused and explicit, and can help reduce the load on your working memory.

The 10 principles of effective skill practice

Here again are the 10 principles of effective practice I gave earlier:

1. Practice the task you need to do.

2. The single most effective learning strategy is retrieval practice.

3. When you practice retrieval, only correct retrievals count.

4. Aim to do at least two correct retrievals in your first study session.

5. Space your retrieval attempts out.

6. Review your learning on a separate occasion at least once.

7. Space your review out.

8. Review at expanding intervals for long-term learning.

9. Interleave your practice with similar material.

10. Allow time for consolidation.

How well do these principles apply to skill practice?

The first principle is self-evidently applicable to skill learning, but clearly so integral to the process of learning a skill that it doesn't need to be explicitly named. Obviously, when you practice a skill you are practicing the task you need to do.

The main difference between information learning and skill learning, in terms of effective principles of practice, is that retrieval is not the focus in skill learning. Automatization is. For motor skills, it's all about automating sequences. For cognitive skills, it's about learning to automatically recognize certain patterns, and automating subprocesses. Practice, then, is far more important for skills (since automatization only occurs through repeated practice), but *retrieval* practice is not — automatization means that the actions will occur automatically in response to actions or environmental cues, rather than requiring a database search.

What this means is that, whereas the central element in effective practice for information learning is retrieval practice, the central element in effective practice for motor skill learning is deliberate practice. For this reason, most of the above principles are not applicable to motor skill learning (I'm specifying motor skills here, because cognitive skills by their nature contain an information component for which retrieval is still relevant). However, those that remain are even more critical — those principles that relate to spacing, interleaving, and reducing interference.

1. Skill level is directly related to the amount of deliberate practice you do.

2. Break the skill down into self-contained sequences or sections that can be practiced separately.

3. Practice difficult sections more than easy ones.

4. Respond to errors by repeating the section/ sequence more slowly and carefully, building up speed on successive repetitions.

5. Focus on only one aspect of the skill at a time.

6. Systematically vary the way in which you perform the skill, and the circumstances in which you do it.

7. Monitor and reflect on your performance.

8. Practice regularly but space it out — be aware of how long you can focus before your concentration fades, and allow time for consolidation.

9. Interleave your practice of sequences or problems with similar, but non-identical, sequences or problems.

10. Use observation of those who are better than you as a means of practice, and of learning new ways.

Putting all this into practice

A brief word about the main obstacles to using effective practice strategies: incorrect beliefs, test anxiety, and failure to set in place the right habits.

Beliefs that stand in the way of effective learning

There's so much research now that makes it clear that spacing and interleaving are far superior to massed / blocked practice, and that testing is far superior to re-reading. Nevertheless, surveys of college students have found that re-reading is the study strategy employed almost all of the time, and teachers continue to teach in blocks, without returning to earlier material until the final exam. Why is that?

A great deal of the problem has to do with the ease of the ineffective strategies. Re-reading is much much easier than testing yourself, and it's easy to fool yourself that you know the material. If you study in a block, you will perform much better on a test at the end of that block than you would if you mixed it up — and the immediate memory is all you know at the time. Spaced, interleaved learning requires (at least initially) faith. But if you try it, you will see the benefits over time.

If you're an older student, practice is even more important, but you may be even less inclined to practice enough. There are several reasons for that. One is that you may remember learning being so much easier, once upon a time. When the same amount of study doesn't produce the same performance, you may be inclined to believe that age has destroyed your ability to learn. While it's true that age can make learning more difficult, it certainly hasn't ruined your learning ability! You may, however, have to get more serious about using effective study strategies than you remember doing when you were young.

The main problem older adults have is interference. Over time we carve ruts in our mind, along the paths we travel a lot, while other paths have time to become neglected and overgrown. We also accumulate more and more information. No surprise that when we learn something new, it has a greater chance of contradicting older information, or getting entangled with similar information (which can be a plus, but isn't always).

Effective practice, and more frequent practice, is thus more important for older adults. In particular, they need to adopt learning strategies designed to counter the problem of interference. Additionally, recent research suggests that older brains may consolidate new memories more slowly. That means you need to support consolidation as much as possible, which means:

- giving yourself time during the day for new information to 'settle' (stabilize)

- reviewing in the evening before bed

- having 'immersion periods', when you immerse yourself in the topic or language or skill

- widely spacing your learning of different skills or topics.

This advice isn't limited to older students! Any student who has trouble consolidating new information — meaning that, although you seem to master the information when first studying it, your memory of it on the following day is poor — will benefit from following these strategies.

Test anxiety

Many students who do poorly in exams suffer from test anxiety. Unfortunately, being anxious makes it more likely that you will indeed perform badly, because your anxious thoughts are using much-needed working memory resources.

Testing yourself is the best way of reducing your anxiety about testing, giving you reassurance that you know the material, and also providing positive test experiences to counteract the negative experiences you've probably had. Unfortunately, those who suffer from test anxiety may be even more reluctant than other students to use testing as a learning strategy. All I can say is that, if you're in this category, you should force yourself to do it anyway! I assure you that you'll soon reap the rewards of doing so.

If you do suffer from test anxiety, I recommend that you ease up on the demands of my standard recommendations. Make sure you don't stretch your spacing too far — it's more important that you find remembering easy. To make up for shortening the review intervals, increase the number of reviews. As time goes on, and your faith in your own abilities increases, you'll be able to stretch your spacing out further.

Remember: if you don't do well when being tested, it's because the interval before the test was too long, and you haven't reviewed enough for the demands of the material. It's *not* because you are 'stupid'! Respond to failure by re-assessing your strategies. Here's a checklist you may find helpful:

- Do you understand the material sufficiently well? If you haven't done this, try drawing a concept map to test your understanding. If the concepts aren't well-connected, or you can't spell out how they're connected, then you need to study the material further (see my book on *Effective notetaking* for more on this).

- Are there particular concepts that you're having trouble remembering? If they're simple concepts, expressed in a word or phrase, try using a keyword mnemonic to help you remember. If they're more complex concepts, try putting the information in a graphical format, such as a picture or diagram or map (I cover this in *Effective notetaking* as well).

- If the material is a skill or has a skill component, break the skill down into its smallest components, and master each component bit by bit.

Habits can break or make you

We might like to think of ourselves as 'higher' beings, self-willed and self-directed, but the truth of the matter is that we are more programmed than we like to think. Our daily lives tend to be ruled by habit, and this is true even if you think you live a free and non-routine life. Habits can be invisible.

There's nothing wrong with any of this! Habits are what allow us to get through our days without having to make a constant stream of decisions. Without habit, it would be as if every day was a brand new day in a brand new place, with new and possibly frightening choices to make. Habits relieve the pressure on working memory, allowing us to focus on what really matters to us.

So, yay for habits. But habits, as we all know only too well, can be good or bad. And in the context of practice, they can either support practice, or they can sabotage it.

If you don't make a deliberate effort to incorporate your practice into your schedule — to turn it into a habit — then the power of habit is not working for you, but against you.

Scheduling your reviews

So don't assume that planning to review means that it will happen! You need to take more concrete action than simply thinking to yourself: "I'll review that on Thursday". When you study, you need to formally take note of the dates on which you'll review the material. If you use a calendar app on your phone or tablet or computer, then put these reviews into your schedule. If you don't, get a calendar with large squares and write down your review dates.

Make sure you note down what material you're reviewing, and which review it is (given that your strategy may change depending on whether it's the first or third or fifth review). If you're planning on long-delayed reviews (as recommended if you want long-lasting learning), then it's particularly important that you write these into your future schedule.

We're all familiar with good intentions. They're easy to keep in the beginning, and even easier to let slide. Writing your intended reviews into your calendar is vital, but may not, in the long-term, be sufficient on its own. You'll find it much more likely that you do those reviews if you build review sessions into your daily routine.

If you're a full-time student, or are studying a language, have a particular time of day in which you do your reviews (with some flexibility, to allow for disruptions to your routine). By that, I mean you review something every day (or nearly so).

If you're only doing a single course, daily review times would be over-kill, but it's even more vital that you establish a regular schedule. If you set your 'review days' for, say, Monday and Thursday, then you'll get into the habit of thinking of them on those days.

Bottom line

There are a lot of details and specific recommendations in this book, that I hope will help you develop an individualized program that maximizes your learning. But if nothing else, I hope you come away with some 'big picture' ideas firmly in your brain and gut — the knowledge that:

- learning is a matter of using appropriate strategies, not a matter of your personal 'ability'

- more than anything else, successful learning reflects effective practice

- there is no one 'right' way to practice that fits all situations and all people

- practice is the key to increasing your effective 'intelligence' (working memory capacity; ability to understand and reason and remember).

Study productively!

Glossary

axon: a long projection extending from the cell body, that carries the output of the neuron away from it.

chunk: a tight cluster of information able to be treated as a single unit when worked with.

code principle: every memory is a selected and edited code, not a recording of real-world events.

cognitive load: the burden on your working memory system made by information-processing tasks.

consolidation: the process of further editing and stabilizing new memories for long-term storage.

context: the information contained in the situation in which you are encoding or retrieving the target information. It includes the physical environment and your own physical, mental and emotional state, as well as information presented at the same time as the target.

context effect: the degree to which the context in which you are trying to retrieve information matches the context in which you originally encoded it affects how easy it is to retrieve.

declarative memory: factual knowledge; information you can make a declaration or statement about. This contrasts with procedural knowledge, knowledge about how to do something.

dendrite: a branched projection of a nerve cell that conducts electrical stimulation *to* the cell body. The name is derived from the Greek word for tree (dendron).

desirable difficulties: a degree of difficulty, such as reading in a hard-to-read font, that encourages learners to put more time and effort into processing the information, resulting in better learning.

distinctiveness principle: memory codes are easier to find when they can be easily distinguished from other related codes.

domino principle: activating one memory code causes other, linked, codes to be activated also.

encoding: the process of transforming information into a memory code, and placing it in your memory.

frequency effect: the more often a code has been retrieved, the easier it becomes to find.

fluid intelligence: cognitive functions associated with general reasoning and problem-solving; often described as executive function, or working memory capacity. This contrasts with crystallized intelligence, which refers to cognitive functions associated with previously acquired knowledge in long-term store.

hindbrain: the brain develops, in utero, in three separate portions, reflecting evolutionary history: the hindbrain, the midbrain, and the forebrain. The hindbrain (the oldest part of the brain) develops into the cerebellum, the pons, and the medulla.

hypercorrection effect: when students are more confident of a wrong answer, they are more likely to remember the right answer if corrected.

interleaving: interspersing practice of one type with practice of other types.

learnable point: important information expressed concisely in a statement that can be easily turned into a question-&-answer format.

matching effect: a memory code is easier to find the more closely the code and retrieval cue match.

metacognition: your understanding of cognitive processing in general, and of your own cognitive processing.

monitoring: strategies to inform you how well you have learned the information in a memory situation so that you can plan your encoding strategies appropriately.

myelin: the (white-ish) sheathing that insulates axons and facilitates speedy communication among neurons.

network: the structure of memory — memory codes that are connected to each other.

neurotransmitter: a messenger chemical in the brain; it is through neurotransmitters that neurons communicate with each other. Examples are GABA, glutamate, acetylcholine, dopamine, serotonin, norepinephrine.

outcome goals: your objective in carrying out a learning task, in terms of the desired outcome. This contrasts with process goals.

priming effect: a memory code is readier to activate, and so easier to access, when memory codes linked to it have been recently activated.

process goals: specific intermediate objectives that need to be achieved on the way to producing the desired outcome of a learning task.

recall: the retrieval of information from long-term memory.

recency effect: a memory code is more readily activated when it has recently been activated.

recognition: the awareness that you have seen or learned this information before. Multi-choice tests assess recognition rather than recall.

reconsolidation: stable memory codes become labile again (capable of being changed) after reactivation, suggesting that consolidation, rather than being a one-time event, occurs repeatedly every time the representation is activated (that is, retrieved from long-term memory).

retrieval cue: something that prompts you to recall a specific memory.

retrieval context: the situation in which you attempt to remember the information. In the study situation, examples include an exam, multi-choice test, classroom discussion, writing an essay, or a brainstorming session.

retrieval practice: the strategy of repeatedly trying to retrieve the information to be learned.

retrieval-induced facilitation: when retrieval practice improves memory for related, untested information.

retrieval-induced forgetting: when retrieval of information blocks the retrieval of other information.

retrieving: finding a memory code; 'remembering'.

schema: a generalized outline or composite framework that has been constructed from a number of specific examples.

spacing: reviewing learning or practicing a skill at spaced intervals, rather than in one concentrated block.

stabilization: the first stage of memory processing, lasting about six hours, during which new information is particularly vulnerable to being lost.

synapse: the place where one neuron makes contact with another; this contact is not physical, but a specialized receptor sensitive to particular neurotransmitters.

working memory: includes the part of memory of which you are conscious; the "active state" of memory. Information is held in working memory during both encoding and retrieval. Working memory governs your ability to understand, to learn new words, to plan and organize yourself, and much more..

working memory capacity: the amount of information you can hold and work with at one time. Now thought to be 3-5 chunks.

References

References for studies cited in each chapter, in order of occurrence.

Chapter 3

Roediger, H. L., & Karpicke, J. D. (2006). Test-enhanced learning: taking memory tests improves long-term retention. *Psychological science, 17(3)*, 249–55.

Atkinson, R. C. (1975). Mnemotechnics in second-language learning. *American Psychologist, 30(8)*, 821–828.

Fritz, C. O., Morris, P. E., Acton, M., Voelkel, A. R., & Etkind, R. (2007). Comparing and Combining Retrieval Practice and the Keyword Mnemonic for Foreign Vocabulary Learning. *Applied Cognitive Psychology, 21*, 499–526.

Pyc, M. a, & Rawson, K. a. (2010). Why testing improves memory: mediator effectiveness hypothesis. *Science (New York, N.Y.), 330(6002)*, 335.

Karpicke, J. D., & Blunt, J. R. (2011). Retrieval practice produces more learning than elaborative studying with concept mapping. *Science, 331(6018)*, 772–5.

Chan, J. C. K., McDermott, K. B., & Roediger, H. L. (2006). Retrieval-induced facilitation: initially nontested material can benefit from prior testing of related material. *Journal of experimental psychology. General, 135(4)*, 553–71.

Kang, S. H. K., Pashler, H., Cepeda, N. J., Rohrer, D., Carpenter, S. K., & Mozer, M. C. (2011). Does incorrect guessing impair fact learning? *Journal of Educational Psychology, 103(1)*, 48–59.

Chapter 4

Vaughn, K. E., & Rawson, K. a. (2011). Diagnosing Criterion-Level Effects on Memory: What Aspects of Memory Are Enhanced by Repeated Retrieval? *Psychological Science, 22(9)*, 1127-31.

Rawson, K. a, & Dunlosky, J. (2011). Optimizing schedules of retrieval practice for durable and efficient learning: How much is enough? *Journal of experimental psychology: General, 140(3)*, 283–302.

Chapter 5

Baddeley, A. D., & Longman, D. J. A. (1978). The Influence of Length and Frequency of Training Session on the Rate of Learning to Type. *Ergonomics, 21(8)*, 627–635.

Cepeda, N. J., Vul, E., Rohrer, D., Wixted, J. T., & Pashler, H. (2008). Spacing effects in learning: a temporal ridgeline of optimal retention. *Psychological Science, 19(11)*, 1095–102.

Cepeda, N. J., Coburn, N., Rohrer, D., Wixted, J. T., Mozer, M. C., & Pashler, H. (2009). Optimizing distributed practice: theoretical analysis and practical implications. *Experimental Psychology, 56(4)*, 236–46.

Chapter 6

Seabrook, R., Brown, G. D. a., & Solity, J. E. (2005). Distributed and massed practice: from laboratory to classroom. *Applied Cognitive Psychology, 19(1)*, 107–122.

Metcalfe, J., Kornell, N., & Son, L. K. (2007). A cognitive-science based programme to enhance study efficacy in a high and low risk setting. *The European Journal of Cognitive Psychology, 19(4-5)*, 743–768.

Richland, L.E., Bjork, R., Finley, J.R. & Linn, M.C. (2005). Linking cognitive science to education: Generation and interleaving effects. *Proceedings of the twenty-seventh annual conference of the cognitive science society. Mahwah, NJ: Erlbaum.*

Taylor, K., & Rohrer, D. (2010). The effects of interleaved practice. *Applied Cognitive Psychology, 24*, 837–848.

Simon, D. a., & Bjork, R. A. (2001). Metacognition in motor learning. *Journal of Experimental Psychology: Learning, Memory, and Cognition, 27(4)*, 907–912.

Kang, S. H. K., & Pashler, H. (2011). Learning Painting Styles: Spacing is Advantageous when it Promotes Discriminative Contrast. *Applied Cognitive Psychology, 26(1)*, 97–103.

Rohrer, D., & Taylor, K. (2007). The shuffling of mathematics practice problems improves learning. *Instructional Science, 35*, 481-498.

Cash, C. D. (2009). Effects of Early and Late Rest Intervals on Performance and Overnight Consolidation of a Keyboard Sequence. *Journal of Research in Music Education, 57(3)*, 252–266.

Dorfberger, S., Adi-Japha, E., & Karni, A. (2007). Reduced Susceptibility to Interference in the Consolidation of Motor Memory before Adolescence. *PLoS ONE, 2(2)*, e240.

Brown, R. M., Robertson, E. M., & Press, D. Z. (2009). Sequence Skill Acquisition and Off-Line Learning in Normal Aging. *PLoS ONE, 4(8)*, e6683.

Nemeth, D., & Janacsek, K. (2010). The Dynamics of Implicit

Skill Consolidation in Young and Elderly Adults. *The Journals of Gerontology Series B: Psychological Sciences and Social Sciences, 66(1)*, 15–22.

Kelley, P., & Whatson, T. (2013). Making long-term memories in minutes: a spaced learning pattern from memory research in education. *Frontiers in Human Neuroscience, 7*, 1–9.

Learning Futures and Monkseaton High School. Spaced Learning: Making memories stick.

Chapter 8

Kornell, N., & Bjork, R. a. (2008). Optimising self-regulated study: the benefits — and costs — of dropping flashcards. *Memory, 16(2)*, 125–36.

Kornell, N. (2009). Optimising Learning Using Flashcards: Spacing Is More Effective Than Cramming. *Applied Cognitive Psychology, 23(9)*, 1297–1317.

Chapter 9

Wulf, G., Shea, C. & Lewthwaite, R. (2010). Motor skill learning and performance: a review of influential factors. *Medical Education, 44*, 75-84.

Ericsson, K.A., Krampe, R.Th. & Tesch-Romer, C. (1993). The role of deliberate practice in the acquisition of expert performance. *Psychological Review, 100*, 363-406.

Ericsson, K.A. (1996). The acquisition of expert performance: An introduction to some of the issues. In K. Anders Ericsson (ed.), *The Road to Excellence: The acquisition of expert performance in the arts and sciences, sports, and games*. Mahwah, NJ: Lawrence Erlbaum.

Starkes, J.L. et al. (1996). Deliberate practice in sports: What is it anyway? In K. Anders Ericsson (ed.), *The road to excellence: The acquisition of expert performance in the arts and sciences, sports and games. Mahwah, NJ: Lawrence Erlbaum.*

Hall, K., Domingues, D., & Cavazos, R. (1994). Contextual interference effects with skilled baseball players. *Perceptual and motor skills, 78,* 835–841.

Allen, S. E. (2012). Memory stabilization and enhancement following music practice. *Psychology of Music, 41(6),* 794–803.

Zimmerman, B.J. & Kitsantas, A. (1997). Developmental phases in self-regulation: Shifting from process goals to outcome goals. *Journal of Educational Psychology, 89,* 29-36.

Hallam, S., Rinta, T., Varvarigou, M., Creech, a., Papageorgi, I., Gomes, T., & Lanipekun, J. (2012). The development of practising strategies in young people. *Psychology of Music, 40(5),* 652–680.

Feltz, D.L. & Landers, D.M. (1983). The effects of mental practice on motor skill learning and performance: A meta-analysis. *Journal of Sport Psychology, 5,* 25-57.

Bernardi, N. F., De Buglio, M., Trimarchi, P. D., Chielli, A., & Bricolo, E. (2013). Mental practice promotes motor anticipation: evidence from skilled music performance. *Frontiers in human neuroscience, 7,* 451.

Charness, N., Krampe, R. & Mayr, U. (1996). The role of practice and coaching in entrepreneurial skill domains: an international comparison of life-span chess skill acquisition. In K. Anders Ericsson (ed.), *The road to excellence: The acquisition of expert performance in the arts and sciences, sports and games. Mahwah, NJ: Lawrence Erlbaum.*

Richman, H.B. et al. (1996). Perceptual and memory processes in the acquisition of expert performance: The EPAM model. In K. Anders Ericsson (ed.), *The road to excellence: The acquisition of expert performance in the arts and sciences, sports and games. Mahwah, NJ: Lawrence Erlbaum.*

van Gog, T. & Rummel, N. (2010). Example-based learning: Integrating cognitive and social-cognitive research perspectives. *Educational Psychology Review, 22,* 155-74.

Hussain, Z., Sekuler, A. B., & Bennett, P. J. (2011). Superior Identification of Familiar Visual Patterns a Year After Learning. *Psychological Science, 22(6),* 724 –730.

CPSIA information can be obtained
at www.ICGtesting.com
Printed in the USA
LVHW080445090819
627064LV00006B/238/P

9 781927 166130